Human Factors Reference Guide for Electronics and Computer Professionals

The McGraw-Hill Engineering Reference Guide Series

This series makes available to professionals and students a wide variety of engineering information and data available in McGraw-Hill's library of highly acclaimed books and publications. The books in the Series are drawn directly from this vast resource of titles. Each one is either a condensation of a single title or a collection of sections culled from several titles. The Project Editors responsible for the books in the Series are highly respected professionals in the engineering areas covered. Each Editor selected only the most relevant and current information available in the McGraw-Hill library, adding further details and commentary where necessary.

Hicks • PLUMBING DESIGN AND INSTALLATION REFERENCE GUIDE

Covers the fundamentals of plumbing design and installation for a wide variety of industrial buildings and structures. Culled by Tyler G. Hicks from several McGraw-Hill books.

Hicks • POWER PLANT EVALUATION AND DESIGN REFERENCE GUIDE

Provides concise evaluation and design information for power plants serving many different needs—utility, industrial, and commercial. Culled by Tyler G. Hicks from several McGraw-Hill books and magazine articles.

Johnson & Jasik • ANTENNA APPLICATIONS REFERENCE GUIDE

Includes practical information and guidelines to antenna applications in all areas of communication. Comprised of one full section of Johnson and Jasik's Antenna Engineering Handbook, *Second Edition. Prepared by Richard C. Johnson.*

Markus and Weston • CLASSIC CIRCUITS REFERENCE GUIDE

Collects in one source hundreds of electronic circuits immediately useful in a wide variety of applications. Culled by Charles D. Weston from Markus's Sourcebook of Electronic Circuits, Electronics Circuits Manual, *and* Guidebook of Electronics Circuits.

Merritt • CIVIL ENGINEERING REFERENCE GUIDE

Offers quick reference to major civil engineering fields: structural design, surveying; geotechnical, environmental, and water engineering. A condensation by Max Kurtz of Merritt's Standard Handbook for Civil Engineers, *Third Edition.*

Woodson • HUMAN FACTORS REFERENCE GUIDE FOR ELECTRONICS AND COMPUTER PROFESSIONALS

Presents all essential data on human factors (ergonomics) relevant to the electronics and computer fields. Compiled by Wesley E. Woodson from his Human Factors Design Handbook.

Woodson • HUMAN FACTORS REFERENCE GUIDE FOR PROCESS PLANTS

Makes available to engineers and specialists all essential data on human factors (ergonomics) relevant to the process industries. Compiled by Nicholas P. Chopey from Woodson's Human Factors Design Handbook.

Human Factors Reference Guide for Electronics and Computer Professionals

Wesley E. Woodson

President, Man Factors, Inc.

McGraw-Hill Book Company

New York St. Louis San Francisco Auckland Bogotá Hamburg
Johannesburg London Madrid Mexico Milan Montreal New Delhi
Panama Paris São Paulo Singapore Sydney Tokyo Toronto

Library of Congress Cataloging-in-Publication Data

Woodson, Wesley E.
  Human factors reference guide for electronics and
computer professionals.

  (Mc-Graw-Hill engineering reference guide series)
  "Derived primarily from materials previously published
in Human factors design handbook (McGraw-Hill, 1981)."
  Bibliography: p.
  Includes index.
  1. Computer engineering—Handbooks, manuals, etc.
2. Human engineering—Handbooks, manuals, etc. I. Title.
II. Series.
TK7885.W64   1987   621.39   86-21068
ISBN 0-07-071766-4

HUMAN FACTORS REFERENCE GUIDE FOR ELECTRONICS AND
COMPUTER PROFESSIONALS

1234567890   SEM/BKP   8932109876

ISBN 0-07-071766-4

D
621.39
WOO

Printed by Semline, Inc. and bound by The Book Press.

Contents

Preface

Content of this book has been derived primarily from materials previously published in *Human Factors Design Handbook* (McGraw-Hill, 1981). The purpose of extracting and compiling the material to create this derivative volume is to make these materials more accessible and attractive to those potential users who may have avoided procurement of the much larger and more expensive volume.

In addition to the direct extractions from the previous publication and the organization of the materials into a format more pertinent to the area of interest, a certain amount of updating has been done to provide information derived from recent research in the area of computer human factors.

Although the title indicates emphasis on the electronics and computer professional's interests and needs, information contained herein is also useful to the individual who may have responsibility for acquiring systems and equipment. In the latter case, it is important to know whether recommendations made by the professional reflect an appreciation of and knowledge about the human factors involved in utilization of electronic/computer systems and equipment.

Human Factors Reference Guide for Electronics and Computer Professionals

Introduction

Advances in electronic and computer technology have been extremely rapid in recent years. Although these advances ostensibly offer considerable potential for handling great amounts of information much more effectively than in the past, it has been apparent that, in some cases, individual hardware users and maintainers have been placed under considerable stress to perform the tasks imposed upon them.

As in the case of many technological advances, the human factors involved are new and sometimes not recognized or understood in the new task setting. Such new technology tends to place the human factors advocate in the position of trying to catch up: human factors specialists must initiate studies to investigate what new equipment systems are doing to operators and find out what the potential problems are—after systems, equipment, and components are already in place and use. Experimental work has to be conducted after the fact, thus placing designers and consumers in the awkward position of considering costly modifications or replacement when they encounter operator problems and user-equipment failures.

In many instances, persons responsible for system and equipment design and/or for simply deciding what equipment to buy take it upon themselves to decide what human factors are important and, more frightening, what to do about them. Their decisions frequently are incorrect and work to the detriment of the individual equipment operator.

CRITICAL HUMAN FACTORS ISSUES

The most critical issues observed by users, as well as human factors specialists working in the computer field, relate to the design of visual displays, control systems, and workplace packaging. In terms of visual display, remarkable improvements have occurred relative to the quality and general capability of display devices: display resolution, reduced distortion, improved color, improved brightness contrast, and minimization of flicker and jitter. However, many problems still persist in terms of display formating, i.e., encoding and simplifying operator control of displayed information.

Considerable advances have occurred in the packaging of visual display terminal (VDT) equipment, as well as in certain support furnishings. Packages and furnishings are more mobile and adjustable, and a few good operator seats are beginning to appear. Although all of these improvements appear to stem from dedicated human factors effort, in some instances it is clear that nonhuman factors specialists have used their own logic or opinions to try to improve the operator-machine interface. Computer-related sales people and their brochures are fond of saying "This equipment is user-friendly." However, they then go on to say that their company offers a 10-volume set of operator instructions, plus a 3-week course to learn to use the equipment.

The purpose of the material in this book, then, is to help the reader understand what he or she must do to make electronic displays readable and understandable, controls manipulatable and compatible with user biomechanical capabilities and limitations, and workplaces geometrically compatible with the expected range of user sizes under a variety of task environments.

Following are some of the most important human factor issues to be addressed either by the designer of future equipment or by persons responsible for selecting and integrating current, off-the-shelf equipment.

a. *Visual displays*—CRT resolution, distortion, jitter, stability; screen size, shape, and position; alphanumeric character form, size, spacing; graphics credibility; color discriminability; display format; error identification and recovery; and data encoding methods.
b. *Controls*—Keyboard layout, location, and position; dynamic control choice, size, shape, location, direction of motion patterns, and accessory keying function type position and operation.
c. *Workplace configuration*—Arrangement and dimensions of components; seating; supporting features such as writing surface; storage; communications; control of ambient lighting.
d. *Environmental control*—Thermal; acoustic; visual.
e. *Maintenance*—Ease of access; mobility of equipment; test, service, and remove/replace interface.

Note: Although concern for health and work stress accompanied initial introduction of VDT operations, studies have shown that in addition to there being no danger from radiation, the CRT does not lead to any more eye strain or stress than a similar eye-related task using typical hard-copy materials. On the other hand, poorly designed CRT displays will produce stress just the same as poorly designed hard-copy materials, and proper lighting control is a major factor in either case.

NEW DESIGN VERSUS OFF-THE-SHELF ADAPTATION

Theoretically, a person responsible for a new design can and should do the best possible job of maximizing the effectiveness of his or her product, which includes making sure the operator-maintainer interface elements are compatible with user needs, capabilities, and limitations. In simple terms, the designer must put himself or herself in the shoes of the user, and design the hardware from that point of view. This will help to avoid many of the current problems that have occurred because the designer was hardware-characteristics oriented (even the labels used often reflect this latter point of view).

Many problems can also occur when equipment is purchased off the shelf and assembled on the basis of personal opinion rather than functionally defined needs. Fortunately, there are numerous equipment and furnishing choices. Individuals responsible for the purchase of equipment make better choices when they are sensitive to the true human factors needs and to differences between equipment that is properly human engineered versus equipment that is not. In many cases this takes no more than simple visual and/or physical examination of a product, for instance, looking at an operational display and observing whether distortions of image or colors are present, manipulating a control device and observing whether it is awkward to use, or sitting at a VDT stand and noticing whether, although adjustable, it is awkward to find and manipulate the adjustment controls from a normal operating position. Perhaps one of the most important things one should realize while becoming acquainted with human factors, is that appearance and decorative aesthetics seldom benefit operability. Avoid being distracted by "designer features."

UNIQUE REQUIREMENTS FOR DESIGN OF ELECTRONICALLY GENERATED, COMPUTER-AIDED CONTROL-DISPLAY SYSTEM TERMINALS

In order for a computer-based, interactive control-display system to be "user friendly" (i.e., not only easy to use but easy to learn to use), certain general human factors oriented conditions must be present. Principal ones are the following.

I. Visual Displays

Visual displays presented on a CRT screen should appear as similar to their hard-copy counterparts as possible. In other words, the electronically produced displays should look like the actual documents they replace (e.g., text, tables, drawings, pictures, lists, etc.) The same design criteria recommended to create optimized hard documents apply equally to electronically produced displays (i.e., criteria pertaining to visibility, legibility, and comprehension). In practice, unfortunately, too many electronically produced displays are created to favor electronic equipment limitations and given programming state of the art or personal opinion.

II. Controls

Computer-interactive controllers (unlike the visual display situation) cannot provide the same one-to-one likeness to provide a hands-on manipulation of hard-copy materials; one cannot physically pick up a document or piece of paper and move it around. This has to be done by substitution or surrogate means. Design of such controls therefore requires consideration of two distinct interface conditions. One is the physical interface between the operator's hand (in some cases foot) and the control; the second is the interface between the controller action and the movement or other action of material displayed on the CRT screen.

III. Off–the–Shelf Component Selection

In most situations, already-designed components are selected and assembled to create a computer-controlled operating system. That is, someone selects VDT component displays and controls, support equipment such as a printer, and the furnishings required to support the components. These are assembled into some kind of work station, and placed in a room where visual, acoustic, and thermal environment is not necessarily tailored to the needs of the VDT operation being planned.

Fortunately there is a wide variety of choices in today's marketplace. If one understands the critical human factors involved in optimizing operator-system interfacing, appropriate display, control, and furnishings components can be found and integrated into an efficient workstation configuration. On the other hand, environmental conditions may require adjustment in order for the operator-workstation interface to be maximally effective.

To select components that have good human engineering features, one should look for the following:

A. VISUAL DISPLAYS—Look for CRT monitors that are the right size for the expected viewing distance, that are free from distortion, that have adequate resolution, that provide proper color separation, and that provide adequate brightness range and contrast control. Alphanumeric character formats must be legible, and the monitor should be packaged so the operator can adjust the screen for best viewing and relief from disturbing reflections. The computer system should provide for a wide range of formating conditions including positioning, spacing, reversing positive-negative and image-background, and shift of displayed elements either by replacement or scrolling. Appropriate cursor and encoding capabilities both for automatic computer control or manual intervention facilities should be available.

B. CONTROLLERS—Look for keyboard packages that are laid out so an operator can easily locate the various controller functions, operate the standard arrays as expected (e.g., typewriting, calculation, dedicated key command, cursor command, etc.), and relate dynamic controller movement and action input to what is shown and/or happening on the display. Make sure that the choice of a dynamic controller device (e.g., light pen, joystick, rolling ball, or "mouse") is appropriate for the particular display-control operation in terms of natural, simple direction of movement relationships, and is free from characteristics that may introduce inadvertent input, limit the precision with which the device can be positioned, or require an operator to use a nonpreferred hand to make precise manipulations.

Some considerations with respect to selecting the appropriate control for efficient manipulation of displayed information items are:

a. *Pushbuttons or keys*—These are suitable for selecting discrete alternatives, initiating predefined commands, entering discrete values, and typing tasks. They generally require the operator to look at them (except for the trained typist) to make sure the right button or key is being selected. They are not a good substitute for properly utilized and designed rotary controls. First, they generally take more space, and second, they require a broader visual scan to locate the proper button within an array. Avoid trying to make all control actions pushbutton or keys. Just because it seems "high tech," or pushing a button seems easy, doesn't mean that such controls are the most task efficient.

b. *Rotary selector controls*—These are excellent for selecting from three to about eight discrete functions. They provide direct visual feedback without requiring special lighting.

c. *Rocker switches*—This type of control is useful for selecting two discrete functions. The switch position provides a visual cue without resorting to special lighting.

d. *Rotary potentiometer controls*—These are effective for making smooth, continuous adjustments for a single dimension through a continuously variable range. They are especially appropriate for such functions as illumination dimming or audio volume control.

e. *Typewriter keyboard (QWERTY)*—The standard keyboard is effective for message entry, especially clear text.

f. *Keypads*—These are appropriate for numerical calculation and/or discrete number code input. Note: The calculator format for numbering is recommended for computer applications.

g. *Joystick-rolling ball*—These devices provide continuous, omnidirectional control of displayed elements that must be moved, positioned, reoriented, etc. They require dedicated space on a work surface, and extra switches must be added to activate a "hook" action. In the case of the joystick, a separate switch can be added on the device; for the ball controller, a switch must be located elsewhere. The joystick is "preferred hand" sensitive; the ball is not, and the control position relative to the operator's body is more critical.

h. *Mouse devices*—These are free-floating, omnidirectional controllers that may work in conjunction with an "electronic tablet" on the work surface area. The manipulation of the mouse requires minimum learning because it allows the operator to move in a natural manipulatory manner. It does not require a dedicated space on the work surface (other than room to maneuver); the mouse device is not attached to the work place except by a small electrical wire lead. Additional switches can be mounted on the device (assuming the package is shaped so the operator can grasp the device and manipulate the switches easily at the same time). Cursor target hooking can be as simple as merely pressing down on the mouse housing. In some applications the mouse can be used over a graphic or picture mounted on a desk in order to enter instructions to modify a similar surrogate display on the CRT screen. A cross-hair element is usually mounted on the mouse package to allow the operator to visually position the controller over a selected starting point on the original picture being copied or adjusted on the screen. The mouse controller can also accommodate a switch to rotate an object on the CRT screen.

C. FURNITURE—In recent years furniture manufacturers have introduced a variety of adjustable VDT support units. The preferred ones incorporate separate height and tilt-swivel adjustments for display monitors; fore-aft, height, and tilt adjustments for keyboards; and height adjustments for various work surface options. When picking one of these units, one should make sure that adjustments will place the displays and controls within the dimensional ranges that will accommodate a full range of operator body sizes, and equally important, that the adjustment controls are located where they can be operated while the operator sits in the normal operating position, i.e., not leaning or slumped.

Similar utility considerations should be sought for support furniture to be used with printers. One should look for units that make it easy to store printer paper, that are easy in terms of threading the paper from storage to the printer (an operator should not have to crawl under the table to thread paper), and that have suitable gathering bins to prevent paper from dumping onto the floor after it comes out of the printer. Such units should be mobile so an operator can place the unit-printer in a convenient and accessible position relative to the primary workstation he or she is using.

Consideration should also be given to other special accessories such as an adjustable copy stand; conveniently placed storage options for reference manuals and a tape machine; and portable, supplementary lighting units that can be fastened to the primary VDT furniture.

Finally (and perhaps as important as anything else), one should select a chair or seat that has been designed to minimize discomfort and allow an operator to position himself or herself in the proper working posture. Key considerations include: (1) seatpan height adjustment so the operator, regardless of size, can keep his or her feet flat on the floor without have the seat cut off leg circulation; (2) backrest that supports the full back up to the shoulders and with adjustable back angle; (3) full swivel seat so the operator can turn from side to side; and (4) castors so the operator can shift the chair from one position to another. Arm rests are preferred when there are no conflicting considerations with the main work table elements. Some convenience considerations are suggested for special operations such as data processor stations, e.g., storage for female operators' purses (lockable units are desirable).

If a work station is to contain more than one video display, the individual displays should be arranged (and/or be mobile) in a quasi semicircle so the operator can, with small head or seat adjustment, look directly at the display (e.g., line of sight perpendicular to screen planes). Finally, consideration should be given to convenient placement of communications equipment such as telephone or intercom.

An important support function consideration in the design of workstations is to make sure that all equipment cords are arranged so that they do not interfere with station task operations or become tripping hazards for operators.

It is suggested that operators of video equipment be given direct control over ambient room lighting where practicable (preferably adjustable dimming). Control for such lighting should be located conveniently "on" the work station so the operator can compare light adjustments with their effect on the CRT display.

When VDTs are set up in rooms where there is considerable natural light from windows, blinds should be provided to cut the glare from these sources. Whenever possible, select rooms without windows. Glare from windows not only impacts directly on the operator's eyes, but produces reflections on video screens. Room surface colors should be medium to dark, but not so dark that a blackout condition exists.

Other equipment in the work room should be placed so that their panel lights do not reflect on workstation displays.

Systems Concept

GENERAL SYSTEMS PRECEPTS

Systems Concept

A "system" as used in this chapter (as opposed to "subsystem" or "component" as used in later chapters) refers to larger, more complex, mission-oriented groupings of elements into an integrated, functional whole. The system typically includes a physical facility, equipment, furnishings, and fixtures and involves a variety of people who use, operate, or maintain it.

System Development

Although not all systems develop from scratch (i.e., some are merely updates or expansions of previous systems), the same general steps are usually taken:

1. Concept formation
2. Preliminary design and evaluation
3. Detailed design
4. Prototype development and test
5. Design modification (if required)
6. Production and delivery (may include preliminary field evaluation before full production)

Human Factors in System Development

Human factors should be considered at the concept formation stage and at all the succeeding development stages. However, at the systems concept level, the principal considerations should include the following:

1. Deciding what roles humans are to be assigned in system operation, i.e., administrative management, operation, maintenance, and general use
2. Deciding where, when, and how humans will interact with subsystems and components, directly or indirectly

3. Deciding what has to be done to provide humans satisfactory living and working environments to ensure not only their safety but also their efficient performance and comfort
4. Determining what human constraints impact on the system design and eventual system performance and deciding how to ensure that humans will not become the weak link in the system

Principal Objectives

The principal objectives should be to design a system that:

1. Is "adapted" to the human, as opposed to creating a system in which the human has to do all the adapting
2. Provides the human in the system the wherewithal to perform in the best manner of which he or she is capable
3. Does not subject the human to extreme physical or mental stress or to possible injury or death as a result of either some equipment malfunction or unpremeditated operator error
4. Provides personal satisfaction for the user in terms of both successful operation and pride of ownership

Determining the Scope of Human Factors Concern in Systems Conceptualization

Avoid the temptation to limit attention to those human factors that pertain to the key operators or maintainers of new equipment. Although an emphasis on primary users may be important during the conceptual phase of system development, the ultimate success of the system operation invariably depends on the effective performance of all subsystems and components, new or old, including all the people who may be involved, directly or indirectly.

1. Potential People Categories That Should Be Considered
Managers and supervisors
Administrators and assistants
Equipment operators
Technical and maintenance personnel
Housekeepers and laborers
Customers and clients
Production personnel
Service personnel
Delivery personnel
Weapons operators
Communications personnel
Visitors and VIPs
First-aid and hygiene personnel
Counselors and training personnel
Security personnel
Disaster control personnel

2. Potential User Characteristics That Should Be Considered
Age
Sex
Cultural background
Training and experience
Size
Mobility
Dexterity
Coordination
Reaction time
Sensory response
Cognitive response
Motor response
Health and handicaps
Physiological tolerance
Motivation
Fatigue limits
Strength
Metabolic requirements
Adaptive limits
Equilibrium
Intelligence level

3. System Features That Should Be Considered

Procedures, instructions, manning, and training for the job, task, or activity

Information input (display)

Information output (control and speech)

Body control and support

Reach and clearance

Demands: speed, accuracy, duration, and strength

Environmental stresses and hazards: temperature, noise, and radiation

Task stresses: complexity and inconvenience

Operational stresses: combat, space, crash, and fire

Consumer acceptance: appearance

Note: The system user is concerned with operational effectiveness, not with the designer's problems of creating the system. In many cases the user does not have the time or inclination to compensate for design deficiencies; the designer does.

ANALYSIS PROCEDURE

General Procedure for Analyzing and Developing the Human Factors Requirements and Solutions in Major Systems Development

1. Mission Analysis

All major systems concepts are developed around some stated mission. The proposed mission should be analyzed in terms of clarifying its purpose and objectives. These will be the underlying basis for all succeeding decisions regarding both the projected hardware and the facility and personnel requirements.

2. Definition of Operational Requirements

Once the general mission purpose and objectives are firmed up and everyone agrees on them, reasonably detailed operating requirements should be defined in order to clarify the demands that will be made on the eventual elements of the system. In the initial stages, these requirements may have to be of a fairly qualitative nature. But in the final stages, requirements should be refined to the point where quantitative values are established for each requirement. These requirements will be the basis for defining functions that have to be performed by physical elements, such as hardware, facilities, or software, and/or by operators, technicians, maintainers, or managers.

3. Definition of Operational Constraints

The general mission should be further studied, and the initial operating requirements should be reviewed with respect to the conditions under which operations must take place; the rules and regulations that the system must adhere to; and the physical, political, social, and economic factors to which the system must be responsive. These constraints, along with the basic operating requirements, provide the basis for defining the functions the system must include and for analyzing alternative means to accomplish the functions.

4. Function Identification

Functions (or processes) should be defined to fulfill each of the requirements identified above. They should be defined in such a way that there is minimal bias regarding how a function will be accomplished. That is, functions should not be described as either "machine" or "human"; rather, there should be independent process descriptions that allow one to consider alternative methods for accomplishing a given process. These descriptions will serve as the basis for examining alternative techniques for performing functions using objective evaluative criteria rather than subjective opinions (which are subject to argument).

5. Function Allocation

Once the various functions are clearly stated and agreed to by all those involved in system concept development and approval, alternative approaches to function accomplishment should be examined. As indicated above, objective evaluation criteria must be established against which to compare alternative function accomplishment methods, modes, or techniques. An important aspect of function allocation is the decision making regarding whether certain functions should be performed by humans rather than by some physical feature, electromechanical device, tool, or software procedure. While it is important during this process to have experts in electromechanical design, structures, materials, and other physical disciplines, it is equally important to have the participation of specialists in the physiological, biological, and psychological disciplines, since these people know more about the capabilities, limitations, and behavioral characteristics of humans as they may be considered to perform particular functions.

6. Definition and Analysis of the Human-System Interface

Once an initial decision has been made regarding which functions humans may be called upon to perform in system operation and use, a list of interface features should be developed in order to analyze the nature of the probable task the humans will be called upon to perform. That is, interface features imply human sensory, cognitive, and motor activity, and it is important to analyze these activities in terms of basic human capacity, temporal and accuracy limits and therefore establish a basis for defining interface design characteristics. All possible interface intersections should be listed, regardless of immediately apparent importance; i.e., not only should the typical control-display interfaces be identified, but also the host of other interfacing features such as furnishings, vehicles, tools, job aids, and life support and habitability elements required to complete the total human operating, living, and survival environment should be identified as well.

The final list developed by this analysis provides the basis for preparing preliminary system, subsystem, and component specifications. However, prior to finalizing these specifications, an investigation should be made of alternative state-of-the-art hardware (as well as other interface elements) to determine *(a)* what may already be available to perform the function (and meet human interface requirements), *(b)* what new devices will have to be developed and designed and/or what modifications of off-the-shelf hardware choices may be required, and *(c)* the requirements for interfacing new and old elements to satisfactorily complete the total system.

7. Preliminary Task Analysis

This is a step that, although typical in military systems development, is often neglected in domestic systems conceptualization. It is important, however, in that it provides the basis for adding appropriate human factors information and/or requirements in the primary hardware specification. Basically, task analysis should focus on creating preliminary descriptions of what humans will do in the system, how they will do it, and what the critical input-output characteristics are between human-machine and operating environment. These descriptions should be encoded in such a way that the earlier functional definitions show a one-to-one relationship with the hardware being used to perform a task. The descriptions should further identify at least the following:

a. The location of the task activity
b. The physical elements associated with the task (e.g., equipment, displays and controls, tools, documents, furnishings, communication devices, internal and external visual and auditory implications, and environmental factors, including ambient illumination, noise, vibration, and acceleration); and the people elements associated with the task (e.g., other operators)
c. Input-output requirements (the type and amount of information)
d. Time and accuracy requirements
e. Potential failure modes and effects (including effects on operator performance and potential hazards)
f. Implications regarding operator skill requirements

Various analysts have devised a number of analytic worksheets for accomplishing task analyses, and a few of these are provided as examples. However, the important thing is to develop one which "fits" the particular system problem and which provides maximum visibility of all the interrelationships that are foreseen and of all the possible problem areas that designers need to address when they go about selecting and/or developing hardware, facilities, and environmental control systems. In other words, the form is not so important as the assurance that all the related factors are being considered.

8. Preliminary Human-System Concept Evaluation

Before a system specification is finalized, it is often desirable to perform a limited concept evaluation. Although this might be done merely by reviewing all the documentation developed during the above steps, some more "visual evaluation" is sometimes helpful.

Typically this is done by most designers when they create artists' sketches, three-dimensional models, and/or preliminary full-scale mockups. Unfortunately, the objective of many of these devices is to "sell." Although this is evaluative in one sense, such devices are frequently not utilized for the more objective analysis of how well the proposed system will work. The principal purpose of the three-dimensional scale model or mockup should be to reexamine certain human-machine-facility interactions in terms of spatial association, activity flow, convenience, ingress and egress, and other factors associated with effective human performance, e.g., visual sight lines, possible noise problems, housekeeping or maintenance problems, and emergency evacuation problems.

Although the model is obviously a useful sales tool, this should not prevent use of the device as a definitive design evaluation tool by the engineering and design staff. Although this may appear obvious, many such devices (originally created by the design staff) have been taken away from the designer by marketing people before they have been fully ex-

ploited to evaluate the proposed system concept.

For certain systems, it may be desirable to create full-scale models to test geometric features prior to finalizing a particular subsystem element, e.g., an operator console, a driver station, or a jetliner buffet or lavatory. Such mockups provide an excellent chance to use the proposed configuration, with "live subjects" performing the expected tasks. Although more of this should be done later, during the predesign phase of system development, it may also be important even at the conceptual stage.

SYSTEM MISSION

Mission Requirements Identification Checklist

When and how will the mission begin and end?

Where will the mission take place?

How will the mission interact with other system missions?

What are the time constraints for the mission?

What are the environmental conditions under which the mission must operate?

What are the sociopolitical conditions or terms under which the mission must operate?

What are the cost constraints within which the system mission must be conceived and operate?

What are the personnel availability constraints within which the system or mission must operate?

What are the ecological constraints within which the system must operate?

What signals must be sensed, detected, and processed and what information must be transferred and stored to meet mission objectives?

What natural and man-made hazards may occur during the mission (e.g., extraterrestrial hazards, undersea hazards, earthquakes, floods, acceleration, explosions, fire, and shock)?

What current systems or subsystems, if any, are to be used as part of the proposed mission or system?

What physical movements of people, equipment, or materials will occur during the mission?

What probable human roles are directly or indirectly associated with the mission (e.g., management, operation, maintenance, and support)?*

*Answers to the above and other pertinent questions should be quantified wherever possible. Although some will require fairly blue-sky estimates at the beginning, reiterative review as other information is generated soon provides a solid basis for system specification. From the human factors viewpoint, these answers establish the basis for identifying critical and noncritical human-system interfaces and the need to perform a special human factors study, particularly if human failure and/or injury may occur during the mission.

HUMAN VERSUS MACHINE

Comparison of Human Capabilities with Machine Alternatives

Human	Machine
Can recognize and use information redundance (pattern) in the real world to simplify complex situations.	Has limited perceptual constancy and is very expensive.
Has high tolerance for ambiguity, uncertainty, and vagueness.	Is highly limited by ambiguity and uncertainty in input.
Can interpret an input signal even when subject to distraction, high noise, or message gap.	Performs well only in a generally clean, noise-free environment.
Is a selecting mechanism and can adjust to sense specific inputs.	Is a fixed sensing mechanism, operating only on that which has been programmed for it.
Has very low absolute thresholds for sensing (e.g., vision, audition, and touch).	To have the same capability, becomes extremely expensive.
Has excellent long-term memory for related events.	To have the same capability, becomes extremely expensive.
Can become highly flexible in terms of task performance.	Is relatively inflexible.
Can improvise and exercise judgment on the basis of long-term memory and recall.	Cannot do this; is best at routine, repetitive functions.
Can perform under transient overload; performance degrades gracefully.	Stops under overload; generally fails all at once.
Can make inductive decisions in novel situations; can generalize.	Has little or no capability for induction or generalization.

(continued)

Comparison of Human Capabilities with Machine Alternatives *(continued)*

Human	Machine
Can modify performance as a function of experience; can "learn to learn."	Is not characterized by trial-and-error behavior.
Can override own actions, should the need arise.	Can do only what it is built to do.
Is reasonably reliable; can add reliability to system performance by selection of alternatives.	Is reliable only at the expense of increased complexity and cost, and then only for routine functions.
Complements the machine, i.e., can use it in spite of design failures, can use it for a different task, or can use it more efficiently than it was designed to be used.	Has no such capability.
Complements the machine by aiding in sensing, extrapolating, decision making, goal setting, monitoring, and evaluating.	Has no capacity for performance different from what was originally designed.
Can acquire and report information incidental to the primary mission.	Cannot do this.
Can perform time-contingency analyses and predict events in unusual situations.	Does very poorly at this.
Is relatively inexpensive for corresponding complexity and is generally in good supply, but must be trained.	Is more limited in terms of complexity and supply by cost and time.
Is light in weight and small in size for function achieved for most situations.	To have functional equivalence of the human, requires more weight, power, and cooling facilities.
Is relatively easy to maintain; demands a minimum of "in-task" extras.	Maintenance problems become disproportionately serious as complexity increases.

Comparison of Human Limitations with Machine Alternatives

Human	Machine
Is a poor monitor of infrequent events or of events which occur frequently over a long period of time.	Can be constructed to reliably detect infrequent events or events which occur frequently over a long period of time.
Has a limited channel capacity.	May have as much channel capacity as can be afforded.
Is subject to coriolis effects, motion sickness, disorientation, etc.	Is not subject to these effects.
Has extremely limited short-term memory for factual material.	May have as much short-term (buffer) memory as can be afforded.
Is not well suited to data coding, amplification, or transformation tasks.	Is well suited to these tasks.
Performance is degraded by fatigue and boredom.	Performance is degraded only by wearing out or by lack of calibration.
Performance is degraded by long duty periods, repetitive tasks, and cramped or unchanged positions.	Is less affected by long duty periods and performs repetitive tasks well; some may be restricted by position.
Becomes saturated quickly in terms of the number of things that can be done and the duration of the effort.	Can do one thing at a time so fast that it seems to do many things at once, for a long period of time.
May introduce errors by misidentification, reintegration, or closure.	Utilizes these processes.
Expectation or cognitive set may lead an operator to see what he or she expects or wants to see.	Does not exercise these processes.
Much human mobility is predicated and based on gravity relationships.	May be built to perform independently of gravity.
Is adversely affected by high g forces.	Is unaffected by g forces.

(continued)

Comparison of Human Limitations with Machine Alternatives (continued)

Human	Machine
Can generate only relatively small forces and cannot exert large forces for very long or very smoothly.	Can generate and exert forces as needed.
Generally requires a review or rehearsal period before making decisions based on items in memory.	Goes directly to stored information for a decision.
When performing a tracking task, requires frequent reprogramming; does best when changes are under 3 rad/s.	Has no such limitations.
Has a built-in response latency of about 200 μs in a go–no-go situation.	Has no response latency.
Is not well adapted to high-speed, accurate search of large volumes of information.	Computers are designed to do just this.
Does not always follow an optimum strategy.	Will always follow the strategy designed into it.
Has physiological, psychological, and ecological needs.	Has only ecological needs.
Is subject to anxiety, which may affect performance efficiency.	Is not subject to this factor.
Is dependent upon the social environment, both present and remembered.	Has no social environment.
Diurnal cycle imposes cyclic degradation of behavior.	The machine cycle may be whatever is desired.
Is subject to stress as a result of interpersonal problems.	Is not affected by such problems.
Great differences exist among unselected individuals.	There are no unselected machines.

TASK DEFINITION

Suggested Steps for Developing Preliminary Task Descriptions during System Conceptualization

Although no two systems will be exactly alike and although the steps may not always be the same, the following list of steps may be useful in generating task descriptions to the level necessary to evaluate conceptual design alternatives:

Analyze initial mission descriptions, purpose, goals, and operational requirements to try to isolate obvious potential human activities. Start by listing mission events in order of their occurrence, from the time the mission begins to the time it is concluded. Although primary interest should be in those activities dealing with the main operating events, consider also all the supporting events and activities.

Create a graphic representation of the sequence of events showing key human and hardware elements, flow of information, materials, mobile components, and people.

Sometimes this is referred to as a "pictogram" or "storyboard" method of activity portrayal. At any rate, the purpose should be to create a pictorial scenario of what is expected to happen during the proposed system operation. The objective should be to make it clear (without verbal explanation) what the system does and how it does this via hardware elements, human elements, and even "unseen elements," e.g., telecommunications, paper transfer, and direct verbal communications.

Analyze each of the principal activity clusters and define the tasks involved in each activity, e.g., monitoring a display, operating a control, and writing or inputting an instruction.

Create a graphic representation of each activity or task event, first to show information flow and then to show operator and equipment tasks. Information flow charts usually begin with a simple graphic showing operation and decision points—how they probably are linked sequentially and with feedback loops. Later these should be refined to differentiate between operations and decisions that are expected to be performed by

humans and those which are to be performed by machines. Reference to human versus machine capabilities should be made as these decisions are being formulated.

Next, create an operational sequence diagram that will show the probable mechanical or manual links between human and hardware elements, the visual and auditory links between human and hardware elements and human and human elements, and any other electrical or physical links between hardware elements.

Prepare a general task description and equipment requirement statement for each activity cluster and create a preliminary operator position identification, e.g., supervisor, radar operator, and vehicle driver.

Preliminary Information-Flow Diagraming Example

The scenario for information flow should first be developed independently of whether the decision and operation points should be via human or machine.

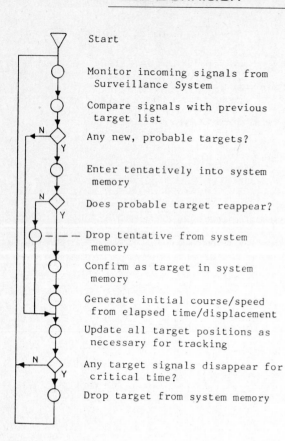

Start

Monitor incoming signals from
Surveillance System

Compare signals with previous
target list

N   Any new, probable targets?

Y

Enter tentatively into system
memory

N   Does probable target reappear?

Y

— | — Drop tentative from system
memory

Confirm as target in system
memory

Generate initial course/speed
from elapsed time/displacement

Update all target positions as
necessary for tracking

N   Any target signals disappear for
Y   critical time?

Drop target from system memory

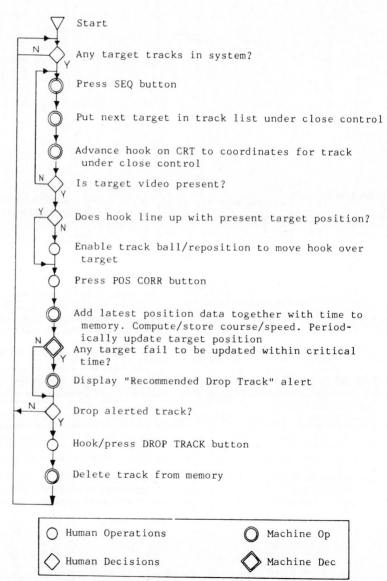

Start

N   Any target tracks in system?

Y

Press SEQ button

Put next target in track list under close control

Advance hook on CRT to coordinates for track
under close control

N   Is target video present?

Y

Y   Does hook line up with present target position?

N

Enable track ball/reposition to move hook over
target

Press POS CORR button

Add latest position data together with time to
memory. Compute/store course/speed. Periodically update target position

N   Any target fail to be updated within critical
Y   time?

Display "Recommended Drop Track" alert

N   Drop alerted track?

Y

Hook/press DROP TRACK button

Delete track from memory

○ Human Operations   ◎ Machine Op

◇ Human Decisions   ◈ Machine Dec

Refined information flow after human and machine assignments are made.

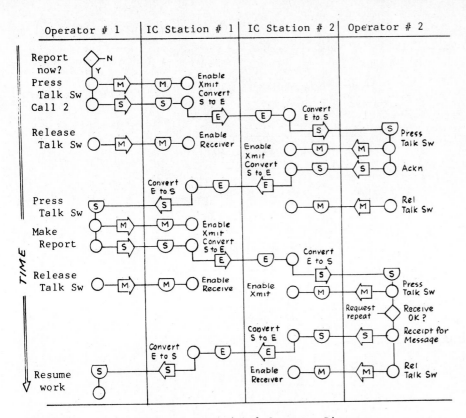

Symbols

◇ Decision
○ Operation
⬠ Transmission
⬡ Receipt
◖ Delay
▢ Inspect/Monitor
▽ Store

Notes on Operational Sequence Diagram

Links

M   Mechanical or Manual
E   Electrical
V   Visual
S   Sound

Station or subsystems are shown by columns. Sequential time progresses down the page.

Hypothetical operation sequence diagram.

DESIGN SPECIFICATIONS

Areas to Consider When Performing Task Analyses and Preparing Preliminary Systems Specifications

1. Functions and Tasks
   a. Are the functions and tasks proposed within the capabilities of operators, maintainers, managers, and associated personnel? Consider requirements for the following functions: sensory and perceptual, motor, decision making, and communication.
   b. Do task characteristics impose excessive demands?
      (1) Task duration (possible fatigue effects)
      (2) Task frequency (possible fatigue effects or boredom)
      (3) Feedback (insufficient guidance)
      (4) Accuracy (too demanding)
      (5) Effect of error (consequence criticality or ability to recover)
      (6) Concurrency effects (interference and overload)

2. The Environment
   a. Events requiring operator response
      (1) Speed of occurrence (too fast)
      (2) Number (too many)
      (3) Persistence (too brief)
      (4) Movement (excessive)
      (5) Intensity (too weak to perceive)
      (6) Patterning (unpredictable)
   b. Physical effects
      Temperature, humidity, ventilation, noise, vibration, illumination, acceleration, etc. (too much or too little or too high or too low)
   c. Mission conditions
      (1) Potential emergencies (ability of the operator to recognize and overcome)
      (2) Response demands (ability of the operator to meet speed and accuracy demands)
      (3) Mission effect criticality (effect of operator error on mission success)

3. Equipment
   a. Display (information requirements)
      (1) Too much to assimilate
      (2) Difficult to perceive, discriminate, and track
      (3) Requires excessive response speed and accuracy
      (4) Excessive memory, interpolation, and extrapolation
   b. Control (response requirements)
      (1) Excessively fine manipulations
      (2) Excessive force required
      (3) Excessive speed required
      (4) Awkward direction of movement
      (5) Excessive range of motion
      (6) Direction-of-motion confusion
      (7) Too many simultaneous movements

4. Supporting Elements
   a. Furnishings
      (1) Seating (inadequate support, adjustment, mobility, or security)
      (2) Writing surface (lack of surface area, insufficient or ill-proportioned surface area, or improper angle or height)
      (3) Storage (lack of storage space, inadequate space, or inconvenient location)
   b. Communications
      (1) Telephone (inadequate number of internal and external connections, telephone sets, and terminal controls)
      (2) Intercom (inappropriate use in terms of contribution to general noise level or because of it)
      (3) Public address (feedback problems and inappropriate use in terms of contribution to general noise level or because of it)
      (4) Radio (inadequate fidelity and volume-level control)
      (5) Signs and signals (inadequate conspicuousness, visibility, or legibility)
      (6) Message delivery (inadequate collection facilities or distribution system)
   c. General illumination, heating, and air conditioning
      (1) General illumination (day and night)
      (2) Supplementary illumination (at specific work stations)
      (3) Emergency lights (in the event of power failure)
      (4) Natural ventilation (air circulation)
      (5) Mechanical heating, cooling, and ventilation (control within comfort range)
   d. Personnel safety
      (1) Restraint systems
      (2) Safety guards
      (3) Personal protective equipment and garments
      (4) Fire extinguishers (general and local)
      (5) First aid (stations and special fixtures)
      (6) Materials and substances (combustion and toxicity)
      (7) Survival equipment

SYSTEMS CONCEPT
Standards

PROGRAM ORGANIZATION

**Organizing for Human Factors
Participation in System Development***

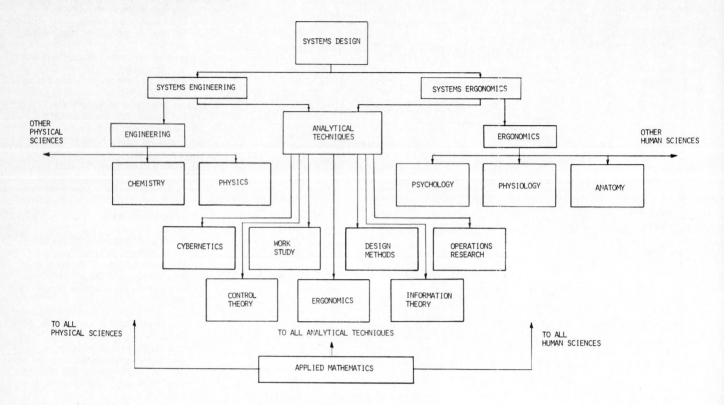

*"Ergonomics," a term used by Europeans and others, is synonomous with "human factors" as used in the United States.

PROJECT STAFFING

Expertise Required for Human Factors Effort during System Conceptualization, Design, Development, and Test

Most system development projects proceed systematically through the following phases: concept formulation, preliminary design, detailed design and development, production, and test and evaluation. The following areas of expertise are typically required during various system development phases.

1. Basic System Experience

A variety of people specialize in human factors research and applied human engineering. They have various educational backgrounds, e.g., anthropometry, medicine, physiology, and psychology. Some are more research-oriented, and others are more design-oriented. For total system development, it is wise to pick someone who is more or less a "generalist," i.e., someone who is not biased toward one discipline (e.g., anthropometry or medicine) or toward one aspect of the system (e.g., life support, controls and displays, or training). Finally, pick someone who has had total system planning and development experience, preferably

with systems akin to the one being planned. This type of person understands both the human factors requirements and the problems of planning, designing, and producing systems within the typical industrial constraints of time and cost.

2. Specialty Experience

Add to the human factors staff and/or employ consultants when it can be determined that the proposed system may have some unique human factors problems. For example:

Space projects require special medical and physiological expertise.
Complex information-processing systems projects require perceptual-motor psychologists.
Systems that must "fit" a wide range of users of different sizes, with different biomechanical capabilities, etc., require anthropometric expertise.

3. Research Experience

Add to the human factors staff and/or hire a consultant or human factors subcontractor when it can be determined that the proposed system development may require special scientific study or experimentation. For example:

A behavioral scientist for sensory, motor, and cognitive studies
An anthropometrist for dimensional surveys of the human body
A training psychologist for skill development studies

STANDARDS

It is generally recognized that the humans in any system can perform better when the interfaces between them and the system are somewhat consistent (or standardized) relative to other, similar systems in which the same individuals may become involved. Although human factors standards have been specially created for certain systems (e.g., the military), there has not yet been similar standardization in other types of systems. The planner or designer can approach the objective of complete standardization to some extent by using various professional engineering and government regulatory standards (although it must be recognized that there will be frequent discrepancies between these in the human-system interface design area). A few of the more useful standards that should be reviewed during system concept formulation are the following:

American National Standards Institute standards: Industry standards developed by the American National Standards Institute, 1430 Broadway, New York, N.Y. 10018. These are often referred to as "ANSI standards."

SAE recommended practices: Society of Automotive Engineers, Inc., 400 Commonwealth Drive, Warrendale, Pa. 15096. These standards relate to automotive, industrial, and similar machine systems.

OSHA standards: Health and safety standards for industry developed by the U.S. Department of Labor, Occupational Safety and Health Administration, Washington, D.C. 20001.

FMVSS standards: Safety standards for transportation systems developed by the U.S. Department of Transportation.

ISO standards: Various standards developed by the International Organization for Standardization are available from the American National Standards Institute (see above).

Other suggested standards are noted under various topical headings in this chapter.

User-Based Design

GENERAL GUIDELINES FOR DEFINING THE HUMAN FACTORS IMPLICATIONS OF VARIOUS PRODUCT DESIGN PROGRAMS

Point of Entry

One does not always have the luxury or benefit of entering every design program at the very beginning; thus the methods and techniques for human engineering must be tailored for each design program. The important considerations are discussed below.

1. Preconceptual Stage

The ideal point of entry for any program is before the customer has made any specific hardware designations. In other words, the design engineer has an opportunity to work with the customer to define the basic operational requirement. In such cases, the design engineer should consider a "total systems analysis program" to systematically define:

a. The mission and operational requirements
b. The functions that are required to accomplish each mission event
c. The performance requirements for each function
d. The allocation of functions to hardware, software, or human elements

2. Conceptual Stage

The next most opportune point of entry for the design engineer is at the "system design concept" stage, where the designer, although given general directions with respect to certain hardware elements, is allowed to make trade-offs relative to how to put various hardware elements together into a (preliminary) base-line system. In such cases, the design engineer not only should consider the analytic steps outlined above but also should include additional analyses to define:

a. The best design approach for accomplishing each hardware functional assignment, i.e., subsystems trade-off studies

b. Preliminary operator, maintainer, and user task descriptions
c. A preliminary manning and training requirements definition
d. Preliminary information-flow and operational sequence diagrams

3. Predesign Stage

The next best point of entry for the design engineer is at the "preliminary design" stage, where the designer, although constrained by preconditions and by the concepts of the customer, still has an opportunity to perform a number of additional design trade-off studies to provide a rationale for possible modification and improvement of the initial concepts. During this stage, the designer not only should review (and possibly modify) the previously noted analyses but also should consider the following:

a. Human-machine mockup studies
b. Human-machine simulation studies
c. Fundamental (but applied) human factors research studies
d. Time-line and link analyses
e. Refined task analysis and task description

4. Detailed Design Stage

In many cases, the engineering designer may not be given the opportunity to perform the preceding preparatory activities, in which case he or she essentially is told what to design and/or what system components to select and adapt into a system or product. Still, at this stage, the designer should consider a limited analysis program that includes all the above analytic steps, but at some lesser degree of refinement. As a minimum, the designer should consider the following:

a. Creating a statement of the system or product purpose
b. Developing a function-flow schematic
c. Creating information-flow and operational sequence diagrams for the critical operating and maintaining operations
d. Performing link analyses for all key operator, maintainer, and user human-equipment interfaces
e. Identifying critical skill requirement specifications that imply the need for operator, maintainer, and user indoctrination and/or training

f. Performing detailed hazards and safety analyses.
g. Creating and evaluating critical human-machine mockups of key operator, maintainer, and user interfaces (control stations, equipment handling, facilities layouts, etc.)

Design Process Documentation

It has become a way of life in modern design to document how and why one has arrived at a specific design concept and configuration. This is necessary because of the number of approval steps required before the design can be released to production, and it may also be an important resource later if questions concerning product liability arise. In any case, it makes good sense to systematically examine and document each step of the design process simply to satisfy one's own desire to create the most cost-effective product. The following design steps are typical of those which should be fully documented.

1. Customer Requirement

The general purpose of the proposed product or system should be clearly defined at the very beginning, before any serious design work is begun. Even though one may have some preconceived design ideas, it is important to "back up" and make sure that the basic purpose for which the product is to be used is clearly understood, both by the customer and by the manufacturer or designer.

2. Mission Definition

The proposed system should be examined in terms of mission or use objectives, wherein "performance" criteria are established. These will eventually be the measure of how successful the design is.

3. Constraints Environment

On the basis of the specific mission or use objectives, one must identify and clarify all the constraints within which the system must operate and/or within which the design must be adjusted, e.g., cost.

4. A Use Scenario

A dynamic, operating scenario should be created as early as possible to illustrate how

17

the mission objectives are going to be accomplished when the system is put into operation. This is particularly desirable for the purpose of communicating the operational concept to non-engineering-oriented clients or to members of management who do not have the time to read complicated written descriptions.

5. Operational Requirements
Detailed performance specifications should be developed that establish quantitative requirements for the sensitivity, accuracy, speed, tolerance, reliability, delivery schedule, and service and maintenance of the system.

6. Function Definition
All the functions and subfunctions required to accomplish the proposed mission should be identified and described. These should include both development and operation or mission functions, since the development may require as much planning as the operation.

7. Function Allocation
Trade-off analyses should be performed to determine the most cost-effective assignment of basic mission functions, i.e., to hardware, software, firmware, and/or human operators or maintainers. More human factors expertise is required at this part of product or system conceptualization than at any other.

8. Detailed Functional Requirements Definition
At this point, hardware functions and human functions are examined both independently and collectively to define the specific requirements that end items must accomplish. It is at this point that human engineering principles should be introduced in order to make sure that the design concepts that may be proposed are going to be compatible with human capabilities and limitations..

9. System, Equipment, and Facilities Design Concept Definition
Preliminary design trade-offs should be made, and preliminary base-line hardware drawings should be prepared, in order to provide a "first look" at the total human-machine system concept. It is important that human factors be given high priority during these trade-off studies. The base-line system provides the first significant hardware review point, at which major decisions will be made concerning whether to proceed with the concept, rework it, or drop it entirely because of technical or economic difficulties.

10. Detailed Design
At this point, development tasks are divided into hardware and personnel development activities; i.e., the engineers and designers pursue the hardware and software side of the development, and human factors specialists pursue the personnel requirements side. However, close interaction is still vital to the successful accomplishment of both activities since the hardware subsystem and the personnel eventually must join to demonstrate a successful match between human and machine to ultimately produce effective system performance.

Human engineering activities during detailed design should include the following:

a. Human engineering design input: Human engineering principles are applied at each stage of design and include critical review of initial design concepts, approval of drawings before release, and evaluation of mockups. Some human factors research may be required when previously established human engineering practices and criteria are insufficient to make appropriate design decisions.

b. Task analysis: As design concepts materialize, operator, maintainer, and user tasks should be described to reflect how, where, and with what operators and maintainers will be required to perform salient operating and maintaining tasks in order to exercise control of the hardware system.

c. Hazards analysis: Although gross analyses should have been made during concept formulation, it is not until design details begin to emerge that one can identify the critical probabilities of certain hazards due to equipment failures, operator failures, or unique interactions between the hardware system and the environment within which it will operate.

d. Time-line analysis: Although the early mission scenario should provide some general description of time relationships, it is not until design details become available that one can estimate the operating and maintaining timing characteristics of system operation. At this point, it becomes evident whether the planned use of the operator or maintainer is feasible, i.e., whether the tasks imposed on the human are within his or her capability to respond or whether the operator or maintainer is apt to be overloaded or underloaded.

e. Link analysis: The link analysis allows one to examine the organization of human tasks in terms of logical grouping and sequencing of activities. It should be done at both the general workplace and the control interface level.

f. Manning requirements analysis: Gross manning requirements are examined during the early planning and conceptual stage of system development. However, it is not until operator and maintainer tasks have been defined in more detail that one has sufficient information to clearly define what kinds of people and how many people will be required to man the system once it is put into operation.

g. Training requirements analysis: Plans for training operators, maintainers, and general users derive from the task descriptions noted above. Training objectives and a training plan should be developed as quickly as possible following detailed design because early development of key training equipment or facilities may be involved.

h. Training aids, equipment, and facilities design: As soon as training requirements are defined, training support items can be defined, including classroom aids, part task trainers, training simulators, special training facilities, and supporting documentation and other materials. As noted above, because it is necessary to provide certain long-lead-time items at the same time that prime hardware is delivered, one may have to perform an initial training requirements analysis prior to having all the design information that normally is generated by the task analyses and hardware human engineering. These items include such things as complex simulators and special training facilities.

i. Technical publications: Certain special publications must be prepared in time to be delivered along with the product and/or slightly before the beginning of training. These include operations and maintenance service manuals (which may become teaching aids). Human engineering attention should be given to the conceptualization and final preparation of these documents so that they will be made maximally helpful to the user. In military system development, it is required that these manuals be field-tested to ensure their use effectiveness.

j. Human test and evaluation: The military requires an early and continuing human engineering test and evaluation program during system development. It considers that human engineering evaluations at each stage of analysis and design are actually tests of the adequacy of the human engineering effort. The final test, however, is one in which all elements of the total system (including hardware, software, firmware, documentation aids, and trained personnel) are tested together as the ultimate measure of total system effectiveness. It is here that the design, procedures, and training are tested to demonstrate whether all development objectives have been properly met.

USER-OPERATOR EXPECTANCY CRITERION CONCEPT

In order for a new system or product to be "user friendly" it must be conceived and designed so that the user or operator is not confused by the interface between machine and operator. That is, when the operator looks at and tries to control the machine, the machine provides the operating cues, the modes of operation, the feedback, and the performance features the operator expects. It is as simple as labeling the product or device so the user recognizes that *this is the equipment or product I want to use*. Of course, in some cases, the shape of the product provides this information. But in other cases, the purpose of a product has intentionally been disguised by the designer, and the resultant package provides no expected cues as to its intended purpose. The same confusion often carries over to the individual features of equipment, making it extremely difficult for the user or operator to find functional features, recognize detailed operating elements, or understand what to do with these elements after finding them.

Therefore, it is important for designers to know what some of the behavioral expectancies are and thus anticipate the manner in which a potential user or operator expects to approach the new design.

GENERAL BEHAVIORAL EXPECTANCIES DATA TABLE FOR DESIGNERS

Behavior Pattern	Safety Implications	Design Considerations
Built-in habits and/or natural behaviors and associations:	Some appearances, operations, and procedures seem more natural than others to people; hence their first impulse is to interpret, operate, or use a device according to these expectations.	If the product is designed so that it either is or appears to be contrary to natural expectations, the user frequently makes mistakes, some of which could lead to hazardous situations.
1. People tend to assume that an object, device, or package is small enough to get hold of and also light enough to pick up.	If an object actually is very heavy, users may suffer severe strain or lose their balance and fall.	Either the user should be warned or the design should include some type of fastening to prevent the object from being lifted.
2. People assume that a vehicle is designed to "fit" their physical characteristics and thus try to adapt to it, whether they can do so effectively or not.	Lack of proper fit often causes users to assume a precarious position or attempt a reach effort that not only makes their situation insecure but also forces them to assume a fatiguing position or apply uncontrolled forces and manipulatory patterns.	The vehicle should be adjustable when it cannot be made to fit all users by a single configuration.
3. People have learned (and therefore expect) that water faucets and liquid valve handles rotate counterclockwise to increase the flow of a liquid, gas, or steam.	If such valves are designed contrary to expectations, when such valves have to be turned off rapidly (in an emergency), individuals frequently will turn them in the wrong direction first and then make a correction—which could be too late.	Valve handles should be marked and mounted according to standard practice. No liquid valve should be incorporated into a design that does not operate according to this standard.
4. Electrically powered control switches are expected to move upward, to the right, or clockwise to effect electrical continuity, i.e. to turn power on.*	If such switches are designed contrary to expectations, people may turn an electrical device on when they think they have turned it off, thus exposing themselves or others to a functioning system which they assume is off.	Design or select electrical switches and control handles that are compatible with the expected directions of motion.
5. People expect that when they move a vehicle control to the right, or clockwise, the vehicle will move in that direction.	Incompatibilities with user expectations result in confusion and often in loss of control.	In the design of vehicles, vehicle control motions should be similar. If the operator must turn around to face a control, the vehicle-control relationships become confusing. Avoid such arrangements if at all possible.
6. People expect that when they actuate a knob or switch that causes an instrument pointer and/or some part of a machine to move, the directions of motion of both the control and the pointer (or part) will be the same.	Incompatibilities between the motion of the instrument pointer (or part) and that of the control element lead to erroneous dial settings and movements of subelements, often causing delays to correct the error and sometimes considerable danger to the user.	Knob and switch motion directions should be consistent with motion compatibility rules, and no control relationship should be used that does not produce a consistent natural response.†
7. People have become conditioned to certain color meanings (e.g., red for "danger," "fire," and "hot"; green for "OK," "go," and "acceptable"; yellow or amber for "caution," "look out for," and "yield"; and blue (in the case of electrical devices) for "cold" or "cool."	Misuse of colors may cause observers to ignore a caution or danger signal or to assume the presence of one that does not actually exist. Either condition creates confusion and delay.	Observe current color-coding standards (which may differ for various product categories).
8. People's attention is drawn to bright and vivid colors, bright lights, loud noises, flashing lights, and repeated and undulating sounds.	People may be distracted by an inadequately selected display and fail to attend to an important task detail, or an important task may not be given attention when needed because the device, signal, or sound was not designed to be conspicuous. When their	Use stimuli of adequate intensity when attention needs stimulation; do *not* use high-intensity stimulation when distractions should be minimized.

(continued)

Behavior Pattern	Safety Implications	Design Considerations
	attention is drawn to a bright light, individuals may disturb their own visual adaptation level to the point where they cannot see necessary visual detail (as when they face oncoming automobile headlights while driving).	
9. Dark, dull colors appear to advance toward the observer; lighter, brighter colors appear to recede.	People may react instinctively (i.e., "duck") to miss a dark object above their heads, even though there is ample clearance.	Observe normal brightness and color impression conventions in selecting colors for interior surfaces.
10. People expect to face auditory signals in order to corroborate and/or reinforce reception.	Placing an auditory warning signal behind an individual could cause the person to displace his or her eye reference as a result of turning the head or body. If this occurs at the wrong moment, it could cause a vehicle operator to miss seeing something in front of the vehicle.	Locate auditory warning signals so that they appear to come from in front of the operator and also locate them close to head height.
11. People assume a relationship between objects on the basis of their spatial proximity (i.e., they assume that things are related somehow when they are located or grouped together).	If a label or control appears to be associated with a display or another control, the individual's inclination is to use that control to effect a modification of the nearby display, whereas it may be the wrong control.	Careful positioning and/or arrangement of controls, displays, labels, or instructions is required to ensure the proper associations.
12. People generally regard products as being safe; that is, they: a. Assume that a system is *ready* to go and hence do not check. b. Grab a handrail assuming that it is strong enough to support them. c. Assume that a system is not on and hence proceed to touch or manipulate without caution. d. Assume that no one else has turned the system on and therefore do not check to make sure.	Although the assumption of "safe" should not be made, the fact that it is tends to preclude users from *thinking* safety, thus reducing the likelihood that they will anticipate possible hazards and make the necessary check.	Try to design the product so that it cannot be used improperly. If that is impractical, provide guards or other means to cause the user to think about the hazards. Finally, if neither of the above is practical, take steps to warn the user by means of labels or instructions, appropriately coded to imply caution or warning.
13. When about to lose their balance or fall, people instinctively reach for and grab the nearest thing.	Sometimes the things most convenient for grasping are not strong enough to support the individual, or they may be hot or slippery or cause an electrical shock.	Configure the design to provide appropriate emergency supports, or else make sure that hazardous components that might be used for support are not within reach.
14. People instinctively use their hands first to test or explore.	Many injuries occur because users touch an object with their hands or fingers, even when it was intended that the task be performed by means of an intermediate device of some kind.	Either make the product suitable for handling or else make it quite clear that its use requires a device supplied to obviate the necessity for using one's hands.
15. People expect a seat height to be at a level they are used to; a stair angle or tread pattern similar to one with which they are familiar and equally spaced in terms of riser height and tread depth; and a handhold or railing of a size and shape that can be gripped securely.	Individuals tend to adjust their posture, weight, mass, and gait to what they expect. When their expectations are not fulfilled, they are apt to lose their balance, take a wrong step, or trip.	Adhere to recognized standards and guidelines so that furnishings and architectural features are consistent with learned use patterns and are compatible with human physical dimensions.

(continued)

Behavior Pattern	Safety Implications	Design Considerations
Lack of knowledge and/or experience: 1. Across the broad consumer population there is a lack of knowledge about, and experience with, the technological features of many products, e.g.: a. Structural integrity b. Mechanical relationships c. Electricity d. Thermal characteristics e. Gaseous and combustible materials and conditions leading to explosion or fire f. Implications regarding center of gravity and balance g. Acceleration, velocity, and inertial factors	Since consumers generally are not prepared either to anticipate or to analyze conditions or the possible results of incompatibilities among physical elements and phenomena, they inadvertently initiate events that lead to improper and hazardous interactions. And because many consumers do not know what makes structures collapse or tip over, what causes materials to ignite or explode, or what inertial forces will do either to others or to themselves, they do not approach the product prepared to deal with these hazards.	Analyze the potentially hazardous interactions between the naive user and the various product elements to make sure that every means possible of precluding misuse has been designed into the product, that misuse has been warned against, and/or that the potential severity of the hazard has been minimized.
2. Most people know very little about their own physical, sensorimotor, or cognitive characteristics or limitations. Thus they do not recognize: a. How they might cause muscle strain, sprains, or skeletal fracture b. How easily they can be thrown off balance c. How vertigo is induced d. How momentum and centrifugal forces affect their balance or ability to reach or apply force e. What is required for them to see effectively and reliably, hear effectively, and feel (touch) accurately f. The importance of adequate sensory feedback in order to control their actions properly or how various internal and external stresses can modify their interpretation of the feedback phenomena	Product users often do not recognize that they are not getting adequate inputs through their sensory channels or that they are reacting to illusory phenomena, upon which they are basing their decisions, rather than to the true informational input.	Learn more about how the human body works, and about its basic characteristics and limitations and then design in such a way as to minimize overstressing, creating illusory stimuli, or creating sensory contradictions and uncertainties.
3. The average consumer is not familiar with much of the technical terminology used for identification	Typical users do not understand the terminology, or they may misinterpret it and thus proceed on an erroneous basis to misuse the product.	Avoid highly technical terms (i.e., use standard nomenclature and abbreviations) and/or test new terms to make sure that they are clear to nontechnically oriented individuals.
Inattention: 1. Inattention is a phenomenon exhibited by everyone at one time or another, either as a natural trait or because of external and internal stresses, fatigue, boredom, or lack of motivation.	Lapse of attention often results in failure to observe a hazardous condition, read a label or instruction correctly, understand an instruction fully, see a warning light, or watch where one's feet, hands, head, or other body and limb components are going and whether they are clearing obstructions.	Anticipate the possibility of inattention in conceptualizing product design; i.e., think of the design in *use* terms. Where the possibility of inattention may be critical to the proper operation and safety of the product-user system, steps must be taken to preclude misuse and/or to provide means for attracting the user's attention.
Distraction: 1. Many people are easily distracted, either by certain aspects of the product's features or by the procedures for its use.	Distraction from the primary product use task may cause users to miss seeing an important element or event, to fail to observe the path of a part of their bodies or the	Analyze the product design and potential use procedures to eliminate distractions and/or to provide a stimulus to maintain user procedural continuity and reliability, or else design to

(continued)

Behavior Pattern	Safety Implications	Design Considerations
	bodily position they have assumed, or to break their train of thought, causing them to omit a step or fail to remember what they have completed.	minimize the possible effects of procedural errors. The environment within which the product will be used must be known, and affective influences must be guarded against.
Hurrying: 1. Most people tend to hurry at one time or another.	When individuals hurry, they tend not to observe or to react carefully; they may not have time to think out or arrive at decisions based on a full analysis of a given situation. Thus, oversights may lead to inadvertent omissions, inappropriate responses, or even inaccuracies in sensorimotor performance.	Recognize the probability that a user will be in a hurry and design the product so that the use procedures help pace the user's response and/or so that the possibility of a hazardous consequence of an error or omission is minimized.
Deliberate risks and shortcuts: 1. Although risk taking and frequent shortcutting of proper procedures are more typical of certain users (e.g., young people), almost everyone takes some risk or shortcut at one time or another.	When a calculated risk cannot be based on a full evaluation of the potential failure modes and consequences, a product use failure probably will occur. Usually the user has neither the knowledge nor the time to make such an evaluation; hence the probability of failure is high.	Anticipate the possible risk-taking and shortcutting behaviors and design the product in such a way as to preclude shortcutting or make the consequences of omissions less serious.
2. Many people refuse to look at and read signs or other visual warnings.	Failure to see or read labels, instructions, or signs leaves the individual uninformed and therefore more prone to error.	Make critical signs and signals conspicuous, legible, visible, and understandable and place them where the user is expected to be looking.
Complacency, neglect, and overconfidence: 1. Because of overconfidence, individuals often proceed without thinking, observing, checking, or reading instructions or labels. 2. Complacency tends to dull people's sensitivity and alertness. 3. Some people fail to exercise caution because of a neglectful and disorganized approach to life.	When people assume that they know how something should be operated, they often make faulty judgments that lead to errors. In some product use situations, alertness is vital to safe performance.	Anticipate possible lack of interest or overconfidence on the part of the user and design in such a way as to alert and stimulate the user to possible critical aspects of product use and operation. The design should be made as simple as possible, and potential hazards should be conspicuously identified and brought to the user's attention. The design should preclude the operational steps' being performed out of sequence.
Forgetting: 1. Although everyone occasionally forgets things, some people by their very nature seem to forget more regularly than others or to forget because of stress or fatigue.	Forgetting a critical step, for whatever reason, can have disastrous consequences. People also can forget about something they ordinarily know is dangerous or potentially injurious.	Anticipate the possibility of forgetfulness on the part of the user and design out critical features where such procedural omissions could lead to injury. Where necessary and possible, the design should provide some technique to monitor whether the user has forgotten a critical step.
Confusion: 1. People are easily confused by things that are unfamiliar.	Lack of familiar features not only lengthens the time required for a user to respond (perhaps making the response too late for safety) but also may freeze the individual into inaction, resulting in an unsafe condition.	Avoid designing products which are so entirely new that nothing about them (shapes, colors, nomenclature, etc.) is familiar to the user.
2. People are easily confused by complexity (in numbers, sizes, shapes, arrangements, concomitant operations, etc.).	Complexity places considerable demand on human perceptual-motor and cognitive processes, and under these circumstances any outside stress could lead to a collapse of the human system response.	Simplify all operator involvements; i.e., complex decisions should be made by the designer of the product.
3. Most people are confused by	Not only will proper identification	Do not assume that people can

(continued)

Behavior Pattern	Safety Implications	Design Considerations
lack of identification of operating interface elements and by the inability to see or find a display or control or to locate the spot where the hand should be placed in order to pick up the device.	improve understanding, but it will also reduce the time necessary to figure out what to do and how to do it. Too much time and/or faulty deductions can lead to misuse or other human errors affecting operator and equipment safety.	recognize product features and understand them without labels or instructions (often labels or instructions are not included because of aesthetic design objectives or cost economies).
4. Many people are unable to perceive mechanical relationships, geometric relationships, or pictorial representations and to interpret these correctly.	If the designer assumes that all people have his or her skill in interpreting relationships and graphics and if the designer uses these to communicate with the typical consumer, the result will be confusion and misinterpretation by the user, resulting in possible misuse.	Avoid using special graphics or depending on relationships that are familiar only to technical people (i.e., design for the nontechnical user).
Lack of manual dexterity, skill, or practice: 1. Many people either lack the innate capacity for dextrous manipulation or control of their bodies and limbs or else have not developed (through practice) the manual skills needed for many tasks involving the fingers, hands, arms, legs, or feet.	When the use of a product (and the ultimate safety of the operation) is highly dependent on manual dexterity or skill and practice, there is a great opportunity for error at one time or another—error that could lead to an unsafe manipulation or operation.	Try to design products in such a way that they do not require great dexterity, precision, or speed or highly sensitive response to various feedbacks. Designers should understand the innate characteristics of the human sensorimotor servosystem and design so that lags in the human system are taken into account.
2. Some manipulatory and other manual skills are especially degraded when the specific human-product relationship is not optimum (e.g., the position of the operator in relation to the task, the extent and direction of movement, and the rate and rate of change of movement).	Increased physical and mental strain reduces the ability to coordinate body, limb, hand, and finger movements and to make precise direction, rate, and force inputs, and it reduces attention and perceptual awareness of errors.	Human-product design relationships should be "natural," convenient, minimally demanding, and not subject to potential misuse.
3. Some manipulatory skills are especially degraded by gloves and clothing.	Clothing may snag and preclude completion of movement; bulk may cause inadvertent contact.	Provide *ample,* not just minimum, clearance between components.
Spurious autonomic responses: 1. Fear, anxiety, and uncertainty often lead to irregular, uncontrolled motor responses. 2. Panic often leads to unexpected responses (e.g., withdrawal, inaction, and violent and uncontrolled input to a device or control). 3. Pain leads to automatic recoil in most cases, although the opposite may be true of small children.	When responses cannot be predicted under these unusual conditions, many undesirable reactions may occur that have adverse consequences in terms of safety; e.g., a driver may freeze to the wheel or accelerator or suddenly exert forces or movements opposite to those which should be exerted, or a person may jump back after receiving an electrical shock, a burn, a sharp cut, or a blow to the body, twisting his or her body suddenly in an effort to recover from a shift in body mass, etc.	Be careful about deciding how users may react to very sudden disturbances in normal situations; these usually do not allow time for thinking, since they are the result of completely involuntary and unpredictable reactions. The design should preclude the possibility or probability of such situations and provide "wiggle room" for the possible reactions so that they do not set off a chain of hazardous events.
Special behaviors: 1. Children: a. Put things into their mouths.	Choking and strangulation.	Make products too large to reach the child's throat.
b. Put their hands and fingers in holes and other openings (including electric outlets).	Entrapment, burns, cuts, shock, electrocution.	Make openings too small and provide covers that children cannot remove.
c. Try to squeeze into or through openings.	Entrapment.	Make opening too small for children to fit through.
d. Climb up on anything they can (e.g., boxes, furniture, and ladders).	Falling.	Eliminate climbing aids, i.e., handles or openings that can be used as a foothold or handhold.

(continued)

Behavior Pattern	Safety Implications	Design Considerations
e. Crawl into boxes, sacks, closets, pipes, etc.	Exposure to hazards within.	Provide secure closures and locks that cannot be operated by children.
f. Pull on cords, clothes, and tablecloths.	Pulling objects on top of child.	Use cordless systems and place electrical outlets where cords will not be within reach.
g. Pick things up and shake them or pound them on other people or other things.	Poking an object into the eyes and breaking windows.	Use soft material that is too heavy to pick up.
h. Pound on all vertical surfaces with their hands (including mirrors and glass doors).	Cuts.	Use shatterproof glass or plastic.
i. Try to open containers, doors, and drawers and to push all such things closed.	Exposure to hazardous substances or medicine; mashing, pinching, or cutting fingers; and dumping contents of a drawer on self.	Use lids that children cannot open, drawer stops, and large, obvious handles.
j. Try to push chairs, stools, and low tables.	Pushing into glass and tipping furniture over on self.	Add sufficient weight and provide proper support and balance.
k. Try to reach for things on the tops of counters, stoves, and pieces of furniture.	Burns, cuts from falling glass objects, and bruises from falling objects.	Increase the height beyond the reach of children and place controls and burners out of their reach.
l. Try to take everything apart, including toys, books, and all kinds of devices.	Swallowing small parts.	Make as indestructible as possible and difficult to disassemble.
m. Try to turn handles on appliances, doors, and other equipment.	Being exposed to hazards outdoors, turning on electrical power or gas, and starting machines, the operation of which is hazardous to the child.	Place handles out of the reach of children and design them so that children cannot operate them (i.e., design handles that require great force, complex coordination, etc.).
n. Try to ride, operate, or drive any kind of vehicle they can climb upon or get into.	A variety of injuries if machine moves.	Provide locks to keep children out.
o. Push all buttons and pull all levers.	Starting an operation that could lead to hazards if uncontrolled.	Provide locks to keep children out, cover push-button panel, and require special secondary movement before lever will move.
p. Put things into any other things that have holes in them.	Equipment malfunction that leads to a variety of hazards.	Cover openings.
q. Like to push things over (to see them fall).	Cuts and bruises.	Design so that weight and/or balance precludes tipping.
r. Like to poke their fingers and other objects into other people's (especially other children's) mouths, ears, and eyes.	Various physical injuries.	No known precaution except to make all objects less pointed.
2. The aged and handicapped:		
a. Shuffle when they walk and therefore trip over things.	Falling and striking objects.	Provide smooth walkway surfaces and mark steps with a highly contrasting color.
b. Hold objects with a weak grip and often unsteadily and therefore tend to drop them.	Cuts while trying to retrieve a broken object.	Use large handles, shatterproof materials.
c. Do not see well or observe carefully and hence bump into, trip and fall over, or fall from irregular surfaces or stairs.	Falling, striking objects, and tumbling on stairs.	Provide good illumination and high contrast at changes of grade.
d. Do not exert a normal effort to control their posture and hence tend to lose their balance easily.	Falling and bumping into objects, receiving injuries on contact.	Make all walking surfaces level, keep all related objects free from sharp corners or edges, and provide strategically located handholds.
e. Do not hear well or listen carefully and hence often are unaware of impending danger indicated by sounds.	Entering a hazardous zone unaware of impending danger, e.g., an approaching vehicle.	Provide visual warning and use an audio warning signal that uniquely penetrates consciousness.
f. Cannot apply as much force as younger or unimpaired persons and hence approach and try to	Misapplication of force, leading to operational hazard and potential injury.	Reduce operating force, optimize control to preclude misuse, and encourage proper use by

(continued)

Behavior Pattern	Safety Implications	Design Considerations
manipulate things in an unconventional manner.		making it the only way in which a product can be used.
g. Have a limited perceptual awareness and are less apt to think problems through; hence they observe less and attend less to what they are doing.	Failure to notice a potential hazard, leading to a variety of consequences.	Maximize perceptual cues (i.e., more contrast and more conspicuous features) and minimize complexity.

*These expectancies are peculiar to Americans. Europeans expect the reverse, i.e., to turn a switch down or to the left in order to turn power on.

†See W. E. Woodson and D. W. Conover, *Human Engineering Guide for Equipment Designers*, University of California Press, Berkeley, 1964.

TYPES OF SPACE

Besides needing enough space in order to move about and perform various tasks, people react to space in a variety of ways. Several researchers have defined the space surrounding the individual in terms of the limits within which people categorically respond (see the accompanying sketch). *Intimate space* is that area in which a person tends not to allow anyone to intrude unless intimate relationships are expected. *Personal space* is that area within which a person allows only selected friends or fellow workers with whom personal discussion is mandatory. *Social space* is that area within which the individual expects to make purely social contacts on a temporary basis. And, finally, *public space* is that area within which the individual does not expect to have direct contact with others. Obviously, the more intimate the spatial relationship becomes, the more people resist intrusion by others. Personal space factors are important in establishing the privacy requirements for architectural design.

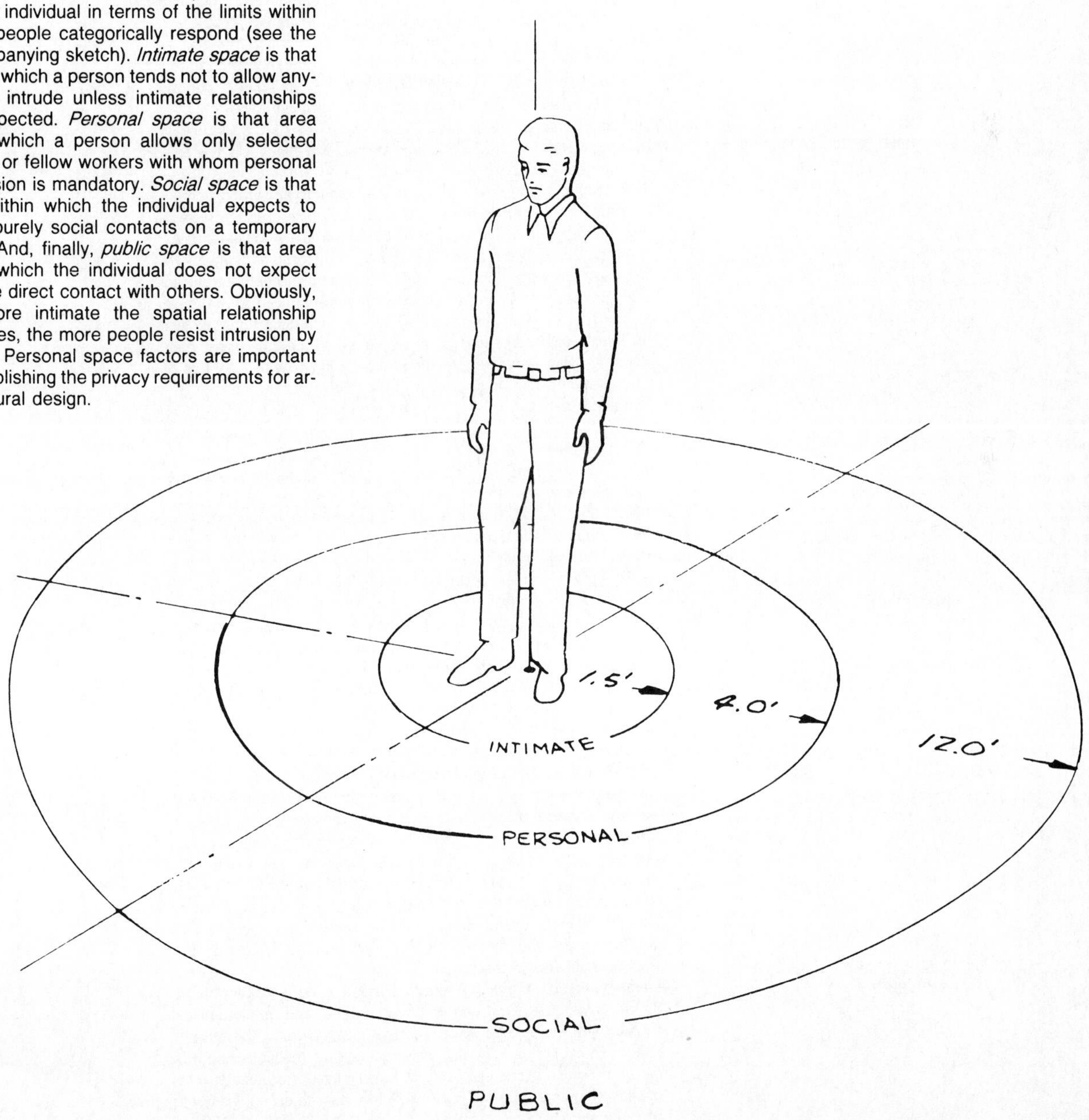

1.5'   4.0'   12.0'

INTIMATE

PERSONAL

SOCIAL

PUBLIC

TYPICAL SUBJECTIVE RESPONSES TO SELECTED SPATIAL FEATURES

Although few research data have been generated with regard to how people respond to specific spatial factors (at least in terms of being able to prescribe precise, quantitative guidelines), it is important for the designer to reflect on potentially negative reactions that often result when a given space is not made compatible with what the user expects in terms of the size, shape, organization, color, and illumination of a particular space. The considerations listed in the accompanying table are suggested as a checklist for the designer.

Space Characteristic	Probable Response
Size (generally volume)	If the space is too small for the number of people, furnishings, equipment, or other objects that occupy it, people will consider it to be crowded. Although they may accept a crowded condition on a temporary basis, they will object to living or working in such a space for extended periods of time. If the space is too large for the people, furnishings, equipment, or other objects that occupy it, people will consider it "unfriendly," inconvenient, and/or overly demanding in terms of communicating, travel distance, maintenance, etc. Although they may accept the "barnlike" atmosphere for temporary periods, they will object to living or working in such a space for extended periods of time.
Shape (generally proportion)	If the space is out of proportion (too narrow, wide, long, high, etc.) for the intended use, people will consider it awkward and often distracting or oppressive. Although they may accept proportional distortion on a short-term basis (i.e., as they pass through briefly), they will object to living or working in such a space for extended periods of time. If the space contains such distortions as all curved surfaces, acute wall junctures, and too many projections or surface changes, people will consider it confusing and difficult to maneuver in and/or furnish. Although they may accept such distortions (or even consider them interesting) on a temporary or one-time basis, they will object to living or working in such a space for extended periods of time. It should also be noted that blind people depend on the constant proportions of right-angle corners to aid them in negotiating a space; such individuals are easily confused by curved surfaces, walls that are not at right angles, and periodic projections that imply they may have reached a turning point. When a ceiling is extremely high relative to the lateral dimension of a space, people feel as though they are working in a pit and that the walls are closing in on them. When a ceiling is extremely low and the space in front of the observer is very long, people feel as though the room is "endless" or as if they will hit their heads unless they duck.
Color and illumination	If a space is dark (unless this is required for a particular operation, such as a motion picture presentation), people tend to become lethargic and less active, or they may feel anxious. As a rule, the less bright a room is, the less cheerful it seems. A small space will seem even smaller. If a space is too bright, people will feel overly exposed, or they will complain of glare or thermal discomfort (even though actual glare in terms of accepted light levels or inappropriate thermal conditions for comfort are not present). If there are too many different colors, too large expanses of very saturated color, or too many and too "busy" patterns of color within a space, most people become irritated after more than a brief exposure to the space. If there is too little color, no visual pattern, or no other decorative "break" in the visual environment, people will find the space monotonous, boring, and eventually irritating to the point of wanting to escape. Although isolated points of highly reflective surface provide interest, all-metallic and highly reflective surface treatments create both subjective and directly objective interference for most people who have to work in the space.
Windows	Generally, most people do not like to live and work in a space that is devoid of windows. First and foremost, people seem to need visual contact with the outside world. Too many windows, on the other hand, can cause the following possible negative reactions: too much glare, too much exposure (fishbowl effect), lack of protection from outside elements, true anxiety (caused by floor-to-ceiling glass at high elevations). *(continued)*

Space Characteristic	Probable Response
Space organization	The internal components within a space and the traffic corridors and entrance and exit locations will seem either well organized or badly organized. The furnishings, partitions, decorative objects, etc., will appear as being either organized or disorganized, depending on the observer's ability to comprehend what things are and where they are with respect to his or her vantage point. Key behavioral response issues are: apparent capability to find one's way to specific locations, apparent ease for interacting and communicating with others with whom the individual must associate, apparent privacy provisions necessary to perform individual tasks. Although these are sometimes conflicting needs, the people who use a space will perform on the basis of how well each of these factors has been executed for *them,* not for the designer or the boss. The organization of internal space components obviously interacts with all the other space characteristics; i.e., the individual perceives and reacts to the combined effects of size, shape, color and illumination, windows, and organization simultaneously. A significant behavioral response will be an individual's interpretation of whether sufficient options are available for local modification of his or her own portion of the space. Even though people may never require a modification, they react to their own space in terms of permanently established restrictions that eventually elicit the feeling that the space is too small, the wrong shape, too dark, or isolated from the rest of the world, for example.
Furnishings	As a general rule, people are sensitive to improperly proportioned furniture, i.e., furniture that is too large, too small, or the wrong shape for the space in which it is placed. Although the designer normally tries to select furnishings that are properly proportioned for the space he or she has created, this may ultimately restrict the efficiency of the individual (e.g., a desk or storage cabinet may be too small). Thus, although the general visual proportions of furniture in relation to space must be taken into account to avoid negative observational responses, shortchanging the individual in terms of specific furniture and use requirements soon stimulates an even stronger negative response.

Unique Responses to the Individual's Location within a Space

Observation and study of what people do when they occupy spaces provide several important thoughts with respect to where a designer expects to place people. For example:

People in auditoriums will not sit next to a side wall, especially if the wall is very high.

People generally dislike sitting facing a wall that is too close, unless they can look out a window.

People generally prefer to sit in a position from which they can observe the entrance to the room and also observe other people, but they do not like to sit so that they are directly observed by others (or think they are).

People dislike having to face other people (in a waiting room, airplane, train, etc.) as they come out of a rest room.

People coming into an auditorium from the rear will tend to take the back seats, not those farther toward the front. Similarly, they will take the seat nearest an aisle, rather than one toward the center of a pew or row of seats.

In a conference situation, most people will take a seat at the corner of the conference table, not an end position or one halfway along the side of the table.

People generally will sit facing a light-colored rather than a dark or highly (saturated) colored or patterned wall.

When there is a row of chairs or stools, most people will seek out one that is not next to one that is occupied.

PSYCHOLOGICAL MOOD

Many attempts have been made to define basic human needs in relationship to designing an environment that is satisfying to the eventual customer, user, or occupant of living and working spaces. The accompanying model illustrates one such attempt to formalize human needs.

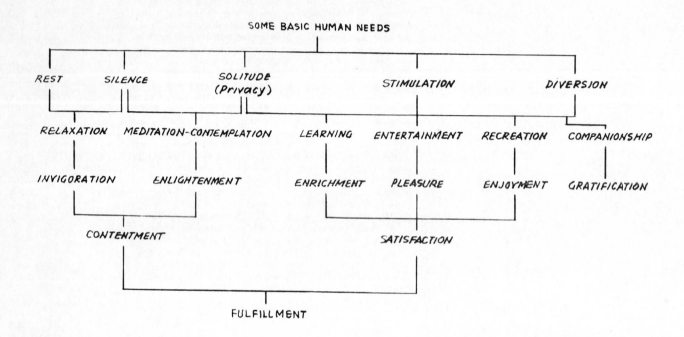

Complete fulfillment probably is beyond the expectation of most people. However, certain negative factors may seriously affect moods, attitudes, and eventual response to the environment. Some of these factors are the following:

Boredom
Apathy
Anger
Fatigue
Depression
Distress
Frustration
Anxiety
Pain

Although any or all of these may not be controllable through design, they can be influenced by it, i.e., by the nature of the task, by the environment within which the task is performed, and by how much stress or lack of stimulation the task involves.

EXPECTANCIES WITH RESPECT TO PRODUCT APPEARANCE

Although there are apparent relationships between various aspects of a product's appearance and the attitudes individuals may express toward it (especially in terms of what people decide to buy), there is little quantitative information to assist the designer in knowing what to do to make a product more accept-able. Key issues regarding product appearance include the following:

Size, shape, and proportion
Color
Texture
Apparent functionality
Compatibility with contemporary concepts (and/or with the style of a particular period, e.g., early American)

It becomes obvious that each of these characteristics hinges upon the particular visual environment within which the product will operate. For example, the size of furnishings needs to be compatible with the size of the space within which the furniture will be placed; the color of the product should be compatible not only with its general surroundings but also with the colors of the other products within these surroundings; the texture of the product should fit with the other textures used in the operating environment; the product must appear to have a purpose within the given environment; and, finally, the product should not appear outdated, unless it is obvious that it is meant to look like an antique, for example.

There is some validity to the theory that an attractive product will be treated with more respect than one that looks "cheap" or poorly designed. However, such concepts are difficult to quantify and write specifications for. On the other hand, it is important to recognize that most people soon neglect products that are hard to maintain in terms of good appearance.

A balance is required between making a product conspicuous or gaudy (to attract the user) and making it so dull and functional that the user tends to neglect it or tire of it after a short period of time.

Almost any product can be made to look attractive *after* its basic functionality has been established. That is, although attractiveness is important at the point of initial purchase, the consumer soon becomes disenchanted if the product is not functional and easy to use.

A product should be attractive from all expected viewing angles; i.e., there should be apparent continuity and completeness from each point of view.

A product should have the appearance of being substantial and sturdy; i.e., it should not appear too fragile and easily damaged.

HUMAN FATIGUE

The problem of fatigue is very important to the consumer, and therefore it should be of concern to the designer; i.e., at least a design should be such that it does not result in activities that are unnecessarily fatiguing in terms of operation or maintenance.

A design may create two different types of fatigue: physiological fatigue, in which the operator's muscles are overstressed, and psychological fatigue (or mental fatigue), which may be caused by design-induced stress, i.e., complexity, high accuracy demands, or environmental implications, such as noise.

MUSCLE FATIGUE

Muscle fatigue occurs because of certain biochemical reactions within the muscle. Nerve impulses acting on a muscle fiber initiate a series of chemical reactions which result from, and contribute to, muscle contractions. Adenosine triphosphate breaks down under the influence of enzymes to create adnosine diphosphate, which in turn releases energy to enable the muscle to perform work. The triphosphate has to be regenerated before another contraction can occur. Energy required for this to happen is supplied by a breakdown of glycogen to lactic acid. Lactic acid is a poisonous by-product that must be removed by oxidation to carbon dioxide and water. The oxygen thus removes the by-products of the energy-producing reaction, which may continue some time after the muscular activity has ceased. The energy for muscular activity thus comes from a reaction which does not depend primarily on the presence of oxygen and which allows work to be done even when the immediate supply of oxygen is insufficient. This often permits the body to make a sudden extreme effort that might otherwise be impossible if the energy always had to be obtained directly from the oxidation of some substance within the muscle fiber.

Oxygen may come either from that stored in red muscle fiber or from the blood supply. As long as the supply is adequate to preclude lactic acid buildup, work is classified as "aerobic." If, however, the rate of work exhausts the reserve of oxygen, the work is said to be "partly anaerobic," and the muscle builds up an oxygen debt; i.e., there is an accumulation of lactic acid in the muscle and bloodstream, causing pain or muscular fatigue.

PSYCHOLOGICAL FATIGUE

The exact nature of psychological or mental fatigue is less clearly understood since it is extremely difficult to isolate and quantify causes and effects. Mental fatigue also confounds the researcher because opposite conditions may produce essentially the same mental or psychological aberration; i.e., one can become just as fatigued by boredom as by overwork. Unfortunately, we know of many instances in which individuals have been able to summon some reserve during a crisis, and thus we tend to believe that, with the proper stimulation, most people can continue mental activity indefinitely. Considerable evidence to the contrary, however, shows us that when an individual works too near mental capacity for long periods, almost any emergency that suddenly occurs may push the individual beyond his or her capacity to cope, the result often being a complete collapse or disorientation. Finally, the question of the threshold of psychological fatigue is further confounded by the fact that an individual's threshold may be stressed to within a few degrees of tolerance by preoperating conditions (prior activities), with the result that he or she has no tolerance to cope with an overly demanding mental task or situation.

Design Implications Relative to Minimizing Potential Muscle Fatigue

Avoid the following:

Designs that require operators to apply near-maximum force capacities over many cycles and for long periods of time

Designs that require continuous, rapid, repetitive muscle contractions for long periods, e.g., pounding, tapping, cranking, or push-pull cycling

Designs that force operators to "hold" some device in a fixed position for long periods without intermittent rest periods

Designs that require operators to maintain an upright posture for long periods without adequate body support (as in the case of a seat)

Designs that require operators to make very long reaches, frequently and for extended periods of time

Designs that require operators to stand or sit in an awkward position and to hold their arms above their heads for a long period of time

Designs that require operators to work in a "bent-over" or squatting position or in a position on their stomachs or backs, with the accompanying stress of holding the head and arms in a strained position

Designs that require operators to bend over and straighten up frequently and over a long period of time

Workplace layouts that require many steps, repeated again and again over a long period of time

Workplace layouts that require operators to sit "askew" (in a twisted position) in order to watch a display and at the same time operate some control (especially a foot control)

Workplace layouts that require operators to hold one foot above a foot control (between pedal depressions) for long periods of time

Workplace layouts that require operators to continuously move their heads from side to side or up and down

Workplace layouts that require operators to step up and down frequently for long periods of time

Tool designs that require operators to hold and push a tool against a work surface or component to maintain contact pressure

Tool designs that require operators to hold a very heavy tool in a precise position for a long period of time

Tool or other equipment designs that require operators to maintain a very tight grip to keep the tool in place (especially if the grip must be maintained for a long period of time)

POSTURE IN RELATION TO FATIGUE

Incorrect posture produces both physical and mental fatigue. As the lower sketches illustrate, the least fatigue occurs when the body can be kept in balance, since that is when there is the least demand on the muscular system to keep the body upright. The musculoskeletal system is most nearly balanced in the standing position. Even in this position, the various sets of muscles require fairly constant flexing to minimize fatigue; i.e., if the body were maintained in a "stiff," erect position, as when a person is at attention, the muscles would be in a state of continuous tension, and thus considerable fatigue would result. On the other hand, a generally erect and balanced position in which the muscles are constantly flexing a slight amount uses up the least amount of energy and produces the minimum amount of fatigue.

When an individual sits, the same principle of muscle balance and minor flexing is desirable. However, in order to create this balance, the body needs to be positioned so that the head mass is easily erected in a generally vertical axis above the torso and buttock masses.

As the accompanying sketches illustrate, the best and least fatiguing posture is one in which it is easy to position the head approximately in a vertical column above the torso and buttock masses (in a seat). If a seat, for example, is sloped too far back, as shown in the second sketch, the individual will naturally pitch his or her head forward to try to reestablish the columnar effect. If the seat is at 90°, as shown in the third sketch, the individual naturally pitches his or her head back to reestablish the columnar effect. Either of the latter conditions produces considerable strain on the neck and back muscles; this is why it is important to provide seats that have the proper seat-pan and seat-back angles.

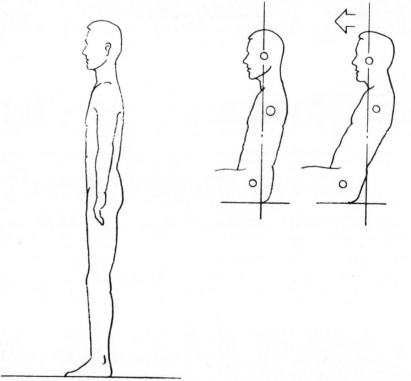

Common Fatigue-Producing Postures

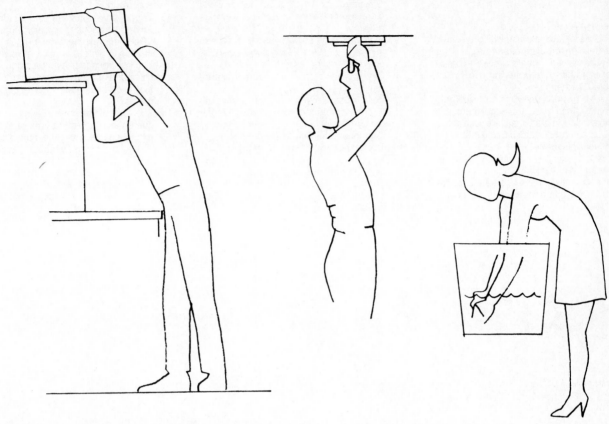

Repetitive lifting to an awkward position. An extended work task located above the worker's head. Bending over for extended periods.

Lack of attention during work-station design planning often results in the worker's being subjected to severe strain, especially if his or her task requires assumption of these poor postures for a considerable length of time. Neural and skeletal disorders often can be traced to the workplaces illustrated.

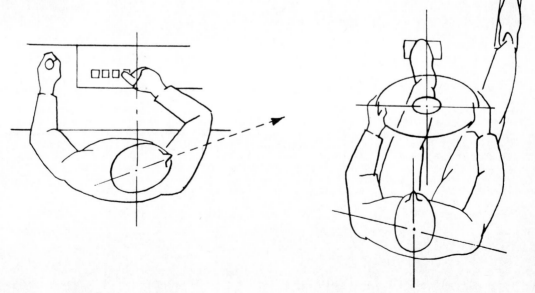

Sitting for long periods in a skewed position.

EQUIPMENT DESIGN AND OPERATING FEATURES KNOWN TO CONTRIBUTE TO MENTAL FATIGUE

1. Too many separate visual displays have to be monitored simultaneously.
2. Visual display formats require extrapolation rather than providing directly usable information.
3. Visual display detail is considerably greater than required; e.g., there are more scale marks than are warranted by the inherent accuracy of the instrumentation.
4. The legibility of visual display details is borderline, requiring unnecessarily close scrutiny in order to detect, recognize, and interpret what is being displayed.
5. Visual displays vibrate because they are not properly shock-mounted.
6. Visual displays are not adequately illuminated, or there are uncontrolled glare sources within the critical viewing envelope.
7. Simultaneous audio communications and/or excessive background noise is present.
8. Overly precise control adjustments are required. ,
9. Poor control dynamics, in terms of force-displacement, control-display, direction-of-motion, and/or movement ratio incompatibilities, are present.
10. There are long delays in informational feedback, i.e., long periods between signals or changes in equipment status.
11. There is a lack of timely indication of whether the equipment is functioning properly.
12. Continuous manual monitoring or control tasks are required that could just as well be automatic, with periodic operator alerting.
13. There is a lack of standardization among various similarly operated pieces of equipment, thus requiring operators to shift their point of reference.
14. The control panel layout is poorly organized, making it necessary for the operator to search for appropriate panel elements.
15. The workplace environment is inadequately controlled in terms of:
    a. Lighting, temperature, humidity, ventilation, noise, vibration, acceleration, pressure, etc.
    b. Support furnishings (standing platform, seating, writing surface, reference storage, restraint system, etc.).
    c. Space, e.g., clearance.

WORK OVERLOAD VERSUS BOREDOM

Mental stress and/or fatigue may be due either to an overload condition or to sheer boredom. Generally it is rather obvious when a design creates an overload condition, i.e., when the operator is required to do too many things at the same time. However, boredom is more difficult to cope with in terms of what can be done to a particular design. It has been suggested that one of the primary methods for dealing with the problem of potential boredom is to make sure that the human is given tasks that are more suitable for his or her unique capabilities, namely, planning and decision making. Alternatively, the machine or equipment should be given those tasks which are more difficult for the human and/or which are monotonous and unchallenging. The following general guidelines are suggested:

To prevent work overload:

Sequence tasks rather than creating overlaps.
Make individual tasks short.
Minimize task precision requirements.

To prevent boredom:

Provide task variety.
Distribute tasks equally throughout the work period.
Assign the operator only significant tasks and make it clearly evident that the operator rather than the machine is in control.

Visual Display Interface

As a general concept, electronically produced displays should look like their counterparts, e.g., hard copy, actual hardware, pictures of drawn graphics. Today's high tech provides an almost limitless capability to do this. One problem exists, however; we do not always try to copy the best model of an item we are displaying. Considerable research has produced guidelines for maximizing the visibility, legibility, readability, and understandability of hard copy. In most cases these guidelines apply equally well to the electronic surrogate on a CRT screen. The characteristics of features on the printed page that make these visible, legible, readable, and understandable are the same for CRT displays. Problems occur when limitations of the CRT are allowed to dictate the characteristics of the display, e.g., lack of resolution; distortion; poor contrast; poorly configured alphanumerics; lack of space between characters, lines, and columns; colors that are not sufficiently different; and so on. In addition, there is a tendency to attempt to put too much on the CRT display at one time, a fault similar to that which publishers perpetrate in hard-copy documents. Similar problems also exist in terms of the ambient visual environment. Glare and surface reflection on the printed page or improper lighting affect how well a person can read from the printed page; these factors also affect how well an operator can read from an electronic display screen.

GENERAL GUIDELINES FOR THE SELECTION AND DESIGN OF VISUAL DISPLAYS

General Principles

1. *Use the simplest display concept* commensurate with the information transfer needs of the operator or observer. The more complex the display, the more time it takes to read and interpret the information provided by the display, and the more apt the observer or operator is to misinterpret the information or fail to use it correctly.
2. *Use the least precise display format* that is commensurate with the readout accuracy actually required and/or the true accuracy that can be generated by the display-generating equipment. Requiring operators to be more precise than necessary only increases their response time, adds to their fatigue or mental stress, and ultimately causes them to make unnecessary errors.
3. *Use the most natural or expected display format* commensurate with the type of information or interpretive response requirements. Unfamiliar formats require additional time to become accustomed to them, and they encourage errors in reading and interpretation as a result of unfamiliarity and interference with habit patterns. When new and unusual formats seem to be needed, consider experimental tests to determine whether such formats are compatible with basic operator capabilities and limitations and/or whether the new format does in fact result in the required performance level.
4. *Use the most effective display technique* for the expected viewing environment and operator viewing conditions (lighting, acceleration, vibration, operator position, mobility restrictions, etc.). Match the display technique to the operator's constraints; do not make the operator match the display.
5. *Optimize the following display features:*

   a. Visibility: Viewing distance in relation to size, viewing angle, absence of parallax and visual occlusion, visual contrast, minimal interference from glare, and adequate illumination
   b. Conspicuousness: Ability to attract attention and distinguishability from background interference and distraction.
   c. Legibility: Pattern discrimination, color and brightness contrast, size, shape, distortion, and illusory aspects.
   d. Interpretability: Meaningfulness to the intended observer within the viewing environment; requirements for interpretation, extrapolation, special learning, and training; and general reliability in terms of retention of meaning

Note: Consideration should be given to possible visual anomalies of the expected user population, i.e., lack of normal visual acuity, color deficiencies, and nearsightedness and farsightedness. These user limitations should be considered not only in terms of selecting the proper display concept but also in terms of executing the specific features that go to make up the display.

Consideration should also be given to creating displays that are easy to read and interpret, as opposed to displays that are merely attractive and/or startling to look at. Avoid the temptation to make displays decorative at the expense of their being readable and interpretable.

When you do not have the choice of creating a display, i.e., when you are required to select ready-made components, carefully study the available products until you find the one that most nearly meets the requirements and guidelines provided in the remaining parts of this section.

Design or select dynamic displays (instruments that have moving elements, warning indicators, etc.) that let the observer know if the display is malfunctioning. If it is not obvious (e.g., a pointer that suddenly drops to zero when it should be indicating some value), consider a special indication of malfunctioning, such as a warning light or flag.

ALPHANUMERICS

Character Size

All things being generally equal (legibility factors, illumination, atmospheric interference, etc.), the larger letters and numbers are, the better they are seen and recognized—up to the point where the observer would be unable to see an entire character at a single glance.

Many people make a common error, however, in applying the above generalization; i.e., they believe that they should also make the character bolder as it is made larger for more distant viewing. As noted in the discussion of stroke width, to maintain legibility while increasing the size of a character, one must also maintain the proper stroke-width relationship.

The accompanying charts provide guidelines for determining how high a letter or numeral has to be for given viewing distances.

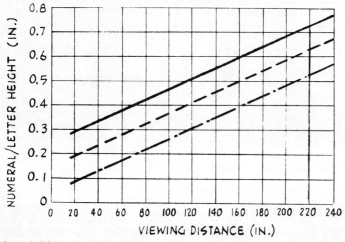
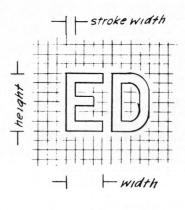

Letter height versus viewing distance and illumination level (minimum space between characters, one stroke width; between words, six stroke widths). (— For instruments where the position of the numerals may vary and the illumination is between 0.03 and 1.0 fL. --- For instruments where the position of the numerals is fixed and the illumination is 0.3–1.0 fL, or where position of the numerals may vary and the illumination exceeds 1.0 fL. –·– For instruments where the position of the numerals is fixed and the illumination is above 1.0 fL.)

Character Shape

The preferred height/width ratio for typical block-type letters and numerals ranges from 5:3 to 3:2, depending on the typeface available.

For emphasis, it is satisfactory to use wider characters with ratios up to 1:1. Avoid selecting characters that are wider than they are tall, however, since such distortion tends to increase recognition time.

LEADERSHIP

LEADERSHIP

When horizontal space is limited, it is permissible to use narrower characters. In some cases, this is even more desirable than trying to crowd the so-called optimum character into a limited space or reducing the space between words.

LEADERSHIP

Avoid selecting extremely narrow characters since they tend to appear blurred and to increase both observation time and errors.

LEADERSHIP

The name of a manufacturer on a product usually is not a critical operator use consideration. A unique typeface here may be an advantage since it clearly separates the name from other operationally critical information.

LEADER

Character Stroke Width

The stroke width of letters and numerals is important in determining how legible the characters will be. If the stroke width is too narrow, there will not be enough definition to make the character clearly recognizable. If the stroke width is too thick, some of the more critical features of the character may be obscured so that it is unrecognizable. For example, the closed portions of a letter such as a B or an R or of a number such as a 6 or 9 may be obscured by a thick stroke width. Similarly, the open space between the horizontal strokes of an F or an E may not be discernible when the stroke width is too thick.

The preferred stroke width for *dark characters* on a light background is 1:6, i.e., the stroke width versus the character height.

The preferred stroke width for *light characters* on a dark background is 1:7.

The rationale for two different stroke widths is based on the fact that light falling on the lighter surface tends to "spill over" into the darker area. In the case of a lighter character, this tends to make the stroke width broaden out; in the opposite case, the stroke width appears narrower. The above recommendations compensate for this apparent artifact of light reflection.

Above all, avoid selecting extremes, either stroke widths that are extremely narrow or ones that are extremely bold.

$\frac{1}{6}$ **LEADERSHIP**

$\frac{1}{7}$ LEADERSHIP

LEADERSHIP

LEADERSHIP

LEGIBILITY

1384

NH ED

AND FOR NOW

WORDS
SPACED

g
L

Because letters and numbers are most often grouped, it is important to recognize that the criteria for a single character's maximum legibility do not always apply. For example, since grouped numbers or letters become too spread out for the observer to grasp the significance of the total pattern, characters should be selected that are taller than they are wide.

The most readable height/width ratio should be somewhere between 5:3 and 3:2.

Letters or numbers within a group should be about one stroke width apart ($\pm$ ½ stroke width).

The space between groups of words or number groups should be about three stroke widths ($\pm$ ½ stroke width).

The space between lines of number groups or words should be at least one-third the height of the tallest character, but in no case should the space between the tallest character of a lower line be less than one stroke width from a character above it that projects below the line.

Figure-Ground Contrast

To ensure adequate legibility, there must be adequate visual contrast between alphanumeric characters and the background against which they are viewed. The following should be considered:

Under normal illumination conditions (i.e., where the observer does not have to be dark-adapted), use dark characters against a light background.

Where the observer must maintain a dark-adapted condition, use light characters on a dark background.

The character should be at least twice as light (or dark) as the background. Greater contrast is preferred and is especially important if the expected observer population includes persons with impaired vision, if the viewing conditions will be less than optimal, or if physiological stress factors are anticipated, such as oscillation vibration, motion, or high *g* forces.

Avoid the use of glossy or highly reflective metallic materials for either the lettering or the background (except where the lettering must be illuminated remotely, e.g., by automobile headlights). Spectral unevenness usually causes portions of the letter or numeral to become illegible as a result of the variation in light reflection.

Suggested Color Contrast Selections in Order of Expected Visual Efficiency

Conditions	Characters	Background
Average or higher levels and quality of illumination	Black	White
	Black	Yellow
	White	Black
	Dark blue	White
	White	Dark red, green, and brown
	Black	Orange
	Dark green and red	White
	White	Dark gray
	Black	Light gray
Poor level and quality of illumination	Black	White
	White	Black
	Black	Yellow
	Dark blue	White
	Black	Orange
	Dark red and green	White
Dark adaptation required	White	Black
	Yellow	Black
	Orange	Black
	Red*	Black
	Blue and green	Black

*Low-level red light is required to maintain the lowest dark adaptation level.

Note: When lettering is internally illuminated so that the letter is always bright against an essentially black background, almost any color is satisfactory as long as the brightness is above the minimal threshold, i.e., as long as the light is visible. Opaque letters must be illuminated with an essentially white light; otherwise, the colored illumination may decrease the contrast between the character and its background.

GUIDELINES FOR OPTIMIZING LETTERS AND NUMBERS TO BE USED ON SIGNS, CONTROL PANELS, AND INSTRUMENT FACES

In spite of some controversy over the best style of character for use in signing, labeling, and instrument marking, at least the following parameters are agreed upon as important to consider in selecting a given character style for a specific application:

1. Proper height/width ratio, e.g., character proportions
2. Proper stroke width, e.g., width of the lines making up the character
3. Maximum figure-ground contrast
4. Adequate overall size for the viewing distance expected
5. A simple rather than an elaborate style, i.e., generally a block style without serifs
6. Upright rather than slanted (italic) characters
7. Adequate intercharacter spacing
8. All capital letters for short labels and signs (as opposed to capital and lower-case letters for extended, instructional materials)

Special Type Fonts

Special type fonts such as those shown have been devised for use on military control panels and visual displays. Individually, these characters are considered maximally legible. They are available in Le Roy Lettering Guide templates and are generally preferred over the standard Le Roy characters.

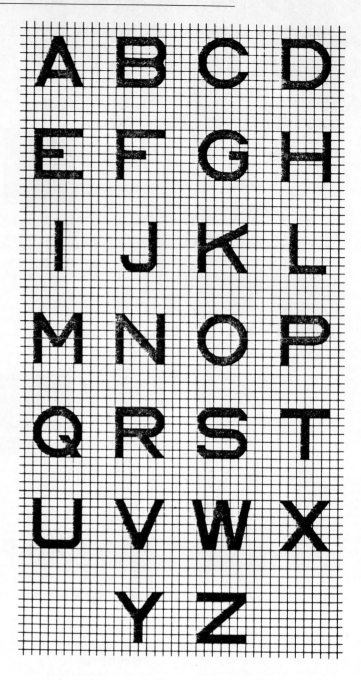

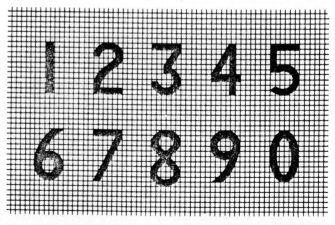

LUMINOUS IMAGE DISPLAYS

Electronically generated image displays such as the CRT (cathode-ray tube) offer maximum flexibility for presenting a wide variety of images and have the advantage of almost instantaneous image modification (as evidenced by the television systems widely used in homes, business, and industry). The following guidelines are offered to assist the designer in making specific choices regarding types of tubes, size, and display mode.

General

CRT-type displays should be viewed from a position perpendicular to the screen; thus the display should be packaged and/or positioned so that the intended observer normally assumes the proper viewing position.

The overall size of the CRT screen should be determined on the basis of the smallest significant detail size that must be visually resolved by the eye at its expected viewing distance.

The screen should be protected from direct high ambient light because it tends to reflect and obscure images and/or excites the phosphor, reducing the inherent contrast of the display.

CRT displays should be packaged so that all adjustment controls for the tube are on the front of the package, i.e., so that they are accessible to both an operator and a maintenance technican. Purely service controls can be placed under a hinged cover to minimize tampering by inexperienced operators.

A transparent safety screen should always be provided over the CRT to prevent radiation or implosion injury.

Except for critical detection (i.e., radar or sonar), CRT displays should not be viewed under complete blackout ambient light conditions.

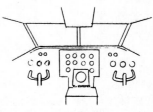

Desirable CRT Performance Characteristics

TUBE SIZE   The diameter of the tube should be a minimum of 7 in (18 cm) for plan position target search operations, 10 to 14 in (25 to 36 cm) for central-area target detection efficiency, and 12 to 19 in (30 to 48 cm) for typical console operator viewing distances (all signal types). Larger tube sizes are acceptable for TV-type viewing and status-board displays that may be viewed by several operators. Sizes larger than 48 in (122 cm) are not recommended because of the loss in brightness and contrast as the viewers are forced to stand further away from the screen.

TARGET SYMBOL SIZE   The minimum size of target pips should be about 12 min of arc measured at the eye, assuming fairly ideal viewing conditions. The preferred size is about 20 min of arc. Alphanumeric symbols should be at least 12 min of arc even with optimum viewing conditions, and about 25 min of arc when conditions are not good (poor viewing angle, high ambient light and glare, etc.).

MINIMUM TARGET SEPARATION   To obtain effective target separation for detection purposes, the minimum separation must be at least 0.1 min of arc.

DISPLAY RESOLUTION   The maximum number of lines possible within the state of the art is always desirable for complex motion displays. Typical TV scan lines and symbolic character height is 10 for adequate character legibility. For digitally generated images that may include handwritten characters, the scan lines and symbolic character height should be about 125 lines per inch.

TV DISPLAY ASPECT RATIO   The standard width/height ratio is 4:3, but ratios of 5:7 or 2:3 provide the greatest legibility.

ACCEPTABLE BANDWIDTH   The acceptable bandwidth is 4.0 to 10 MHz.

GRAY LEVEL   There should be a minimum of five levels for TV; a single level (black and white) is acceptable for most digitally generated images.

DISPLAY BRIGHTNESS   A line brightness of 50 fL ($+$ 40) is required under normal ambient light levels; there should be lower adjustable brightness for lower ambient light conditions.

BRIGHTNESS CONTRAST   The contrast ratio should be as near 90 percent as is practicable.

PHOSPHOR PERSISTENCE   The absolute minimum for target detection tasks is 0.1 s (2 to 3 s is much preferred). As a rule of thumb, images should not persist beyond the time necessary for the eye to detect the presence of a target; prolonged persistence only confuses the image. P-4 and P-7 phosphors are suggested for alphanumeric and/or discrete image displays that change frequently.

FLICKER   Display pulse rates should be compatible with critical flicker frequency response of the eye (CFF); i.e., the particular phosphor-driver combination should not generate pulses in the 30- to 55-Hz range.

GEOMETRIC DISTORTION   Displacement of any image element should not exceed 2 percent of the image height.

COLOR MISREGISTRATION   For additive color systems, the maximum acceptable misregistration is $+$ 65 percent.

PPI DISPLAY SWEEP RATE   Any rate be-

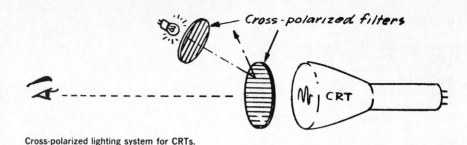

Cross-polarized lighting system for CRTs.

tween 1 and 70 rpm is acceptable, but target detectability is enhanced at the slower rates.

ALPHANUMERIC CHARACTER WIDTH/HEIGHT RATIO   A ratio of 2:3 to 3:5 is best for maximum legibility and fastest recognition.

ALPHANUMERIC CHARACTER STROKE WIDTH   Ratios of 2:6 to 1:10 are recommended.

ALPHANUMERIC CHARACTER STYLE   The closer the character style is to the character legibility guidelines noted elsewhere for printed matter, the more readable the character is. However, when rapid readout is not an absolute requirement, modified, matrix-type characters are acceptable. Although sloping (italic) characters are frequently used and are acceptable for single values (i.e., when there is at least one character-height separation between character lines), vertical characters are preferred.

DISPLAY FORMATS   Display formats should be compatible with the manner in which the display image is generated; i.e., a plan position indicator sweeps an arc, and therefore the shape of the display should provide a round or circular face, and a rectangular face should be used for horizontal or vertically swept display systems, used typically for TV motion pictures and/or alphanumeric status or message boards. For image previewing, a scrolling technique is suggested so that an operator (e.g., a bank proof operator) can key in information from one image while viewing the next.

VIEWING DISTANCE   The optimum for the typical console situation is 18 to 20 in (45.7 to 50.8 cm), e.g., a 12- to 19-in (30.5- to 48.3-cm) screen. A maximum of about 20 ft (6.0 m) should be planned as the limit, depending on the size of the display.

VIEWING ANGLE   This should not be less than 30° from the perpendicular axis. For seated operator console operation, the operator's sight line should not exceed about 30° vertically or horizontally, i.e., about 15° on either side of the operator's normal center-of-the-display viewing angle.

ADJUSTABILITY PROVISIONS   As a minimum, capability should be provided for the operator to adjust CRT brightness and contrast. In addition, however, it is also desirable to provide capability for operator tilt adjustment of the CRT or CRT cover face so that the operator can remove disturbing reflections and glare from the display.

POSITIVE VS NEGATIVE IMAGE   A positive image (i.e., a dark image on a light background) is preferred when the display is used under normal or bright ambient lighting conditions. A negative image (i.e., a light image against a dark background) is preferred under dimout or blackout ambient lighting conditions,

since it is desirable to reduce the amount of light that could affect the observer's dark adaptation level.

COLOR   For normal ambient lighting conditions, use a phosphor color that is near white in color (bluish white, blue, and green are also acceptable, however). For dimout or blackout ambient lighting conditions, use an orange or reddish phosphor to minimize reduction in the observer's dark adaptation level.

When a filter cover is used over the display tube, use a neutral density filter to minimize distortion of the color from the CRT.

Under dimout ambient lighting conditions where reflections may be a serious problem, use a circularized Polaroid filter or a cross-polarized filter-light system to prevent reflections and/or tube phosphor excitement.

By placing the light filter 90° to the CRT filter, it is possible to provide illumination for panel controls without reducing the brightness contrast of the CRT signal.

Multicolored CRT displays are suggested when it is desirable to present typical TV movies, target display encoding for easier differentiation (e.g., enemy versus friendly targets), or other graphic information when it is helpful to differentiate between scales, curves, callouts, etc. Red, green, blue, and yellow are the most reliably differentiated colors.

General Checklist for CRT Displays*

Whenever possible, mount the scope face perpendicular to the operator's normal line of sight. If the operator is standing, his or her line of sight is about 5° downward; seated, it is 15 to 20° downward. Ideally, the operator's line of sight should be perpendicular to the tube center.

Recommended viewing distance is 16 to 20 in (40.6 to 50.8 cm).

Tube size should be proportional to the required resolution. Scopes as small as 2 to 5 in (5.1 to 12.7 cm) in diameter may be used for special cases, e.g., infrequent calibration or tuning.

Scopes with a 5- to 7-in (12.7 to 17.8-cm) diameter may be used when they adequately display the necessary information and when no plotting is required.

For plotting, the scope diameter should be 10 in (25.4 cm) or larger.

The shape of the CRT should be compatible with the information format, i.e., round for plan position displays (PPI) and rectangular for A-scan presentations, document images, etc.

*Human Factors Engineering Design for Army Material, MIL-HDBK-769, 1975.

For a CRT displaying synthetic video, as opposed to raw or unprocessed video, the system should be capable of presenting a symbol of the required size (at least 20 min of arc) for any return resolvable by the system. For synthetic video, in which signals are clipped for digital presentation to the discrete points resulting from digital conversion, the number of points per inch is

$$\frac{\text{Number of points per tube diameter}}{\text{Tube diameter in inches}}$$

For a 14-in (36-cm) screen at 1024 points per tube diameter,

$$\frac{1024}{14} = 73 \text{ points per inch}$$

Symbols should be drawn as point-to-point representations. Symmetrical symbols require an even number of points. For the screen described above (with a viewing distance of 16 in, 41 cm), a 20-min visual angle would subtend 0.093 in (0.236 cm), or 6.77 points. A symmetrical square would have sides extending over 8 CRT points, or ⅑ in (0.28 cm). The area in space which is masked by a symbol equals

$$\frac{\text{Number of points on symbol diameter}}{\text{Number of points on screen diameter}} \times$$

range of coverage

If the screen described above covers 100 km (62 mi) the area masked by the symbol is

$$\frac{8}{1024} \times 100 \text{ km} = 0.782 \text{ km}$$

If the symbol is accompanied by identifying alphanumeric information, each character covers an area representing an area 0.782 × 0.782 km² (0.289 mi²) in the real world. If the scale size is changed to magnify the area covered by the CRT, the area masked by the symbol decreases. For 4× scale (25-km, 16-mi coverage), the area masked by a single symbol is

$$\frac{0.750}{4} \times 0.188 \text{ km}^2$$

Control devices used in conjunction with CRTs (to permit an operator to request information or direct action against a target) should provide a centering accuracy of approximately ±4 CRT points. All such devices (e.g., joy sticks, tracking balls, "stiff sticks," and light pens) require a switch action to indicate that the operator's action is directed to the area currently designated.

A light pen (or pointing device) recognizes only bright line structures specifically called by the computer program and falling within a field of approximately one symbol diameter. To avoid problems, the light-pennable character should not be blink-coded, two individually pennable characters must be separated by at least one symbol, and the cord connecting the pen to the equipment should be attached at a point where it will cause minimal interference with the display viewing and will not accidentally catch on console features.

Summary of Coding Method for Symbols

Code	Number of Steps in Code	Evaluation	
Alphanumerics	Unlimited	Excellent	High information-handling rate; unlimited number of coding steps
Geometrics	20 or more	Excellent	Certain shapes easily recognized; many coding steps
Color	4	Excellent	Difficulty in techniques of reproducing for CRT; objects easily and quickly identified
Blink	2	Poor	Distracting and fatiguing; interacts poorly with other codes; best for attention getting; few steps in code
Brightness	2	Poor	Limited number of steps; fatiguing; detrimental to decoding performance
Line lengths	4	Fair	Limited number of steps; clutters displays
Angular orientation	12	Fair	95 percent of estimates correct within 15 percent
Inclination	24 or more	Fair	Many coding steps, especially with combinations
Visual number (dots)	6	Fair	Few steps
Combinations	Unlimited	Good	Avoid overloading symbols with too much information; complex combination can degrade decoding speed and accuracy

Point resolution is conventionally a fixed percentage of display size. On a 12-in (30-cm) CRT, resolution is about 85 points per inch (1023 × 1023 per display surface). A 4-ft (1.2-m) display of the same matrix would have a resolution of about 21 points per inch. This has little effect on alphanumeric data but puts a definite limitation on graphic displays. On such a 4-ft (1.2-m) screen, lines or points must be separated by 0.05 in (0.13 cm). In both CRT and large-screen displays, line thickness is also proportional to display area.

A 12-in (30-cm) CRT is contained within a visual angle of 41° for a 16-in (41-cm) viewing distance. An operator is able to scan 8.6 in (22 cm) (+15°) comfortably without head movement. The same viewing angle is available at 64 in (163 cm) on a 4-ft (1.2-m) screen. If the larger screen is viewed at a greater distance, the total angular view is decreased; however, element sizes must be increased. With a 16-ft (4.9-m) viewing distance, a 4-ft (1.2-m) screen can be viewed comfortably by a small working group; however, for detail resolution comparable to that of a 12-in (30-cm) CRT, the larger screen can represent only 4 in (10.2 cm) of the CRT display. A choice must be made between the area which can be shown and the amount of detail that can be presented.

Ambient illuminance should not contribute more than 25 percent of screen brightness through diffuse reflection and/or phosphor excitation. The ambient illuminance in the CRT area should have appropriate intensity and color with respect to other visual tasks, e.g., setting controls, reading instruments, inspecting maps, and performing various maintenance and housekeeping tasks, but it should not interfere with the visibility of signals on the CRT display.

The luminance range of surfaces immediately adjacent to scopes should be between 10 and 100 percent of the screen background luminance. With the exception of emergency indicators, light sources in the immediately surrounding area should not be brighter than scope signals.

Viewing hoods or glare-reduction devices should be provided when very faint signals must be detected, and/or a suitable light control filter system should be used to maintain maximum ratios of target signal to CRT background luminance.

Electrically or optically generated displays should conform to MIL-STD-884. Radar CRTs display information more effectively when a symbolic shape code is used in association with, or in place of, raw video pips. The following encoding principles are suggested: The number of discrete symbols should be held to a minimum. Symbol meanings should be standardized.

Symbols should be recognizable without reference to some comparison standard.

If combination codes are used, the respective meanings should be decodable separately without confusion. Two codes maximum combination.

Symbol shapes should reflect a natural relationship to the event or type of target the symbol is supposed to represent.

Codes should be easy to learn and retain.

Symbols should differ widely in shape; avoid variations of a single form, e.g., a circle and an ellipse.

Redundant cues (i.e., those which associate a symbol with a particular size) increase coding efficiency.

External modifiers surrounding a symbol should be avoided.

Reserve blinking coding for emergency situations. There should be no more than two blink rates.

Two brightnesses are the maximum for brightness coding.

Four line lengths can be identified with minimum error.

No more than three blip sizes should be used.

Video Signal	Vertical Resolution		Horizontal Resolution	
	Center (Lines)	Corner (Lines)	Center (Lines)	Corner (Lines)
Monochrome	400	400	800	700
Red, green, or blue	400	400	800	700

Typical TV Equipment Standards*

Television signal characteristics should be based on a 525-line scanning standard, interlaced 2:1, with 60 fields and 30 frames per second. A video bandwidth of 4.5 MHz should be used for color picture signal transmission. A video bandwidth of 10 MHz should be used for monochrome picture signals.

Color Picture Monitors

SCAN SIZE   The normal scan should provide a display in which all four corners of the raster are visible. Width and height controls should have sufficient range to vary the raster size from −10 to +20 percent while maintaining specified linearity.
RESOLUTION   The limiting horizontal and vertical resolution of the monitor should be as specified in the above table for both composite and noncomposite operation.
ASPECT RATIO   The width/height ratio of the raster should be 4:3.

Monochrome Large-Screen TV Projector

SCAN SIZE   Width and height controls should have sufficient range to vary raster size from −10 to +20 percent of the nominal dimensions.
ASPECT RATIO   The width/height aspect ratio of the normal picture should be 4:3.
RESOLUTION   The limiting horizontal resolution with the specified brightness level should be at least 800 lines at the picture center and 700 lines at the corners. Vertical resolution should be at least 400 lines, attained when a composite or noncomposite monochrome video signal is applied to the projector input.
BRIGHTNESS   The projector should provide brightness levels as shown in the table below. Variation in any area of the screen plane should not exceed ±30 percent of the center of the plane.

Monochrome Television Project

Image Size	Screen Brightness (fL)
6 × 8 ft	62
9 × 12 ft	28
12 × 16 ft	16
15 × 20 ft	10
24 × 32 ft	4

Note: 1.0 ft = 0.3 m.

*Defense Communications Agency, Television Technical Characteristics, vol. I, Picture Generation and Display Equipment, AD 805-174, November 1965.

GRAY-SCALE REPRODUCTION   Nine shades of gray and the white background should be distinguishable.
GEOMETRIC DISTORTION   The combined effects of all distortion should not displace any point on the projected display from its correct position more than 1.0 percent of the picture height.
TRAPEZOIDAL DISTORTION   The projector should be capable of correcting "keystone" or trapezoidal distortion resulting from vertical tilt of the screen, i.e., screen tilt from the perpendicular to the optical axis of the projector within $\pm 15°$.
INTERLACE   Displacement of any scanning line from a center position between lines of the alternate field should not exceed 10 percent of the distance between the lines of the alternate field.
WHITE BALANCE   The TV monitor should be capable of producing, from a monochrome input signal, a white that corresponds to CIE illuminant "C" ($x = 0.310$, $y = 0.316$).

TV Quality Variations

At resolutions of 8, 10, and 12 lines, the quality of TV equipment has little significant effect on the accuracy or speed with which standard alphanumerics can be read. At 6 lines, readability is good with high-quality TV (i.e., a minimum of 945 lines).

For group viewing of TV, a minimum vertical resolution of 15 lines/character height is recommended when small visual angles are involved. At 15 lines/character height to total display height is 1/33, and 16 rows of characters can be put on the screen (i.e., as long as the screen visual angle is within 8 min of arc).

For a symbol resolution of 15 lines, the recommended maximum viewing distances for various TV monitor sizes are as follows:

27-in (69-cm) monitor—18 ft (5.5 m)
24-in (61-cm) monitor—15 ft (4.6 m)
21-in (53-cm) monitor—13 ft (3.9 m)
17-in (43-cm) monitor—11 ft (3.4 m)

For pictorial TV viewing, the recommended maximum viewing distances are as follows:

21- to 30-in (53- to 76-cm) monitor—20 to 30 ft (6.1 to 9.2 m)
19- to 23-in (48- to 58-cm) monitor—10 to 20 ft (3.1 to 6.1 m)
17- to 19-in (43- to 48-cm) monitor—6 to 10 ft (1.8 to 3.1 m)
15- to 17-in (38- to 43-cm) monitor—30 in to 6 ft (76 cm to 1.8 m)
9-in (23-cm) monitor—18 to 30 in (46 to 76 cm)

Effect of TV Surround Brightness

The mean value of surround brightness preferred by viewers of broadcast TV is shown in the accompanying graph (plotted for three surround areas at each of five values of peak screen luminance).

Ambient illumination glare should be minimized; i.e., sources should not be located within 60° of the observer's central visual field.

The optimum brightness distribution on the surface of the TV display is approximately 17 fL measured from the central axis and not less than 13 fL measured at the largest angle of view off center. Assuming an ambient light level of 1 fc on the display, this permits viewing in about 10:1 contrast for symbols with regard to the background.

Brightness ratios required for comfortable viewing of large-screen displays are determined by locating two values: (1) the minimum ratio required for adequate viewing and (2) the maximum measure of brightness without annoying aftereffects. The maximum brightness for group displays should not be more than 35 fL. Higher brightness may produce afterimages if the display is viewed for extended periods. An increase in brightness over the 15- to 35-fL maximum contributes no significant improvement in visual acuity.

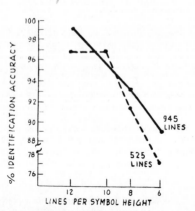

Accuracy of symbol identification for good-quality versus low-cost TV.

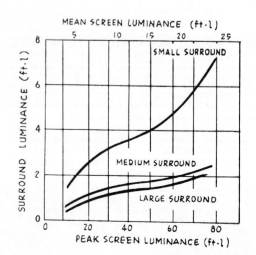

QUANTITATIVE READOUT DIGITAL DISPLAYS

The digital readout generally provides the most rapid and accurate method for presenting purely quantitative information. In selecting or designing such displays, consider the following:

1. The numeral characters should be as legible as possible in terms of character style, height/width ratio, stroke width, and figure contrast.
2. Mechanical counter readouts (drum type) should increase in value by an upward movement of the drum. The face of each drum (and its numerals) should not be buried deeply in the counter window. Individual drums that separate numerals are not desirable because they tend to obscure the numbers and make the total set of numbers harder to read. Wide separators between drums are similarly undesirable. However, such separators are useful for imprinting decimal points and/or commas when the numerical value contains many characters, e.g., 1,000,370.
3. Numbers should always read left to right; vertical arrays are undesirable.
4. Whenever possible, select display systems in which you can use optimized characters (including the drum type, folding character types, etc.). However, when it is necessary to use other types because of a requirement for more rapid response, dot and line matrices are permissible. The accompanying sketches illustrate line or bar matrices that can be used to create effective numeric characters. Although sloping matrix characters have been used extensively, these are less desirable. The slope should never exceed about 11°. Electrically pulsed counter rates should not exceed 50 per second.
5. Although LED (light-emitting diode) displays are effective for use in well-controlled ambient light conditions and/or for nighttime applications, LCDs (liquid crystal displays) are more effective for most general applications because they are easier to see under all ambient light conditions.

6. Dot matrix characters are generally more readable than other matrix-type displays if there is a sufficient number of dots to create the critical features of the harder-to-read characters, i.e., 2, 3, 4, 5, 6, 8, and 9. The 9 × 7 matrix shown in the accompanying illustration is the best format. Note that, although only full steps are required for the vertical dimension, half steps are required horizontally in order to provide proper diagonal strokes for numbers like the 4 that is illustrated.*
7. Electronically generated digital characters should be as free of brightness variations, jitter, and flicker as possible.
8. Although various colors are acceptable when they have special meaning, the preferred colors are black on white and white on black for general use (and/or a bluish color if the display medium is a CRT). In the latter case, avoid long-persistence phosphors, which may produce ghost images.

*A number of recent studies have indicated that, although general recommendations for the design of alphanumeric characters suggest that properly designed solid or continuous-stroke characters provide more error-free, rapid readout, matrix-type numerical characters are just as good, if not better, in some cases. Examination of such studies indicates that these conclusions are influenced by the fact that the displays studied typically do not control all the critical variables to draw a general conclusion that would contradict the previous study results. From a practical point of view, therefore, almost any type of digital character generation technique can be made to produce an adequately legible readout if the design of the characters is such that a sharp, evenly lighted character is produced that contains the critical pattern features that make it discernible from another character that tends to look like it.

Numbers should "snap" into position - should not follow one another faster than 2/sec. to be read.

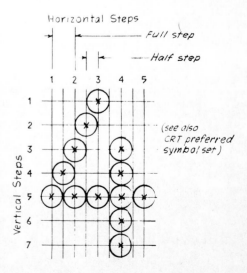

Horizontal Steps
Full step
Half step

Vertical Steps

(see also CRT preferred symbol set)

CRT Symbol Characteristics

DOT MOSAIC The coarsest acceptable mosaic is 5 × 7, as shown in the accompanying illustration. Even these may be marginal compared with larger matrices. Only 35 decoded lines are required. If only numbers and a limited number of symbols are required, such as horizontal or vertical lines or crosses, the number of dots may be further reduced to 35.

STROKE MOSAIC Bar, stroke, or segment patterned symbols can be created as shown in the accompanying illustration. Such matrices are suitable only for numerical characters, not letters. Readability is not as good as with dot matrices if speed and accuracy are required. The seven-segment format is the minimum configuration that is acceptable. Additional segments can be added to improve the legibility of characters. The character format can be slanted, but legibility suffers if the slope is more than 5–7°.

Dot matrices are more readable than segmented types because of the capability to produce curved portions of characters.

SYMBOL SPACING At low brightness levels (1 fL), spacing should be 25 percent of character dimension; at high brightness levels, spacing should approach 200 percent for maximum readability of a single character. For typical applications (e.g., clear text messages or grouped numbers), use 50 percent of average character width for between-character spacing within a word or number group and 75 to 100 percent spacing between words or groups. Spacing between lines of characters should be at least 50 percent of character height.

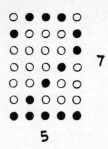

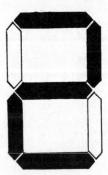

Dot matrices are more readable than segmented types because of the capability to produce apparent curved portions of characters.

A 5 × 7 matrix of dot characters is minimally acceptable; a 7 × 9 matrix, as shown in the illustration on page 545, is preferred.

A B C D E F G H I J K L M
N O P Q R S T U V W X Y Z
1 2 3 4 5 6 7 8 9 0

Minimum character height	3.1–4.2 mm
Maximum character height for	
5 × 7 dot matrix	4.5 mm
Width/height ratio	3:4–4:5
Stroke/height ratio	1:8–1:6
Minimum number of raster lines	10 lines

**Suggested CRT Character Optimized
Set Using a 7 × 9 Matrix**

The suggested character formation shown in
the accompanying illustration contains most of
the alphanumerics and supporting symbols required for data processing. They are designed
to minimize recognition errors, or at least possible confusion between one character and
another. Although lowercase letters have
been designed, they are not recommended
because it is difficult to create lowercase features that are not easily confused, especially
if the observer has to read information rapidly.

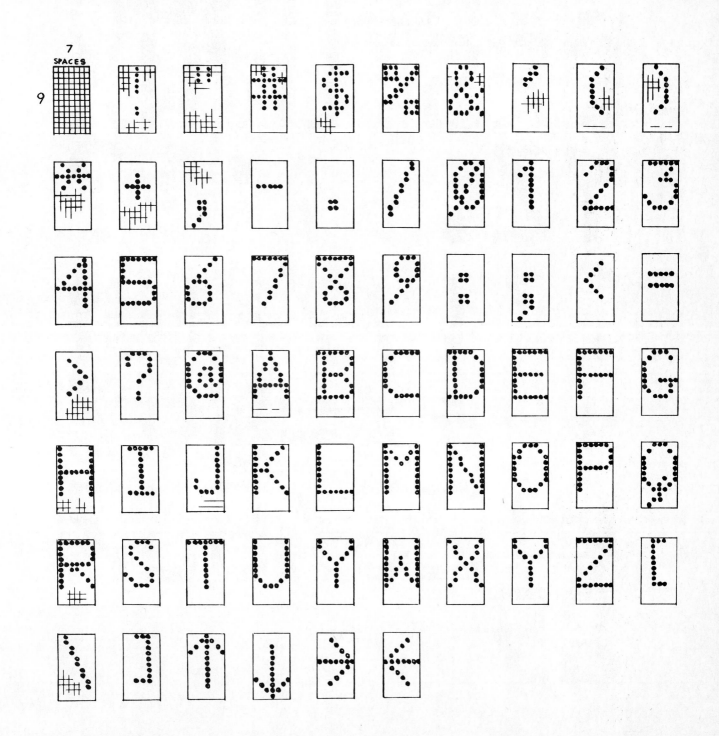

VISUAL DISPLAY INTERFACE
Quantitative Readout Digital Displays

ABCDEFGH
IJKLMNOP
QRSTUVWX
YZ.,;:!?
"'()/&@#
+-÷×=$¢%
0123456
89

Example of an 8 × 14 matrix character set.

General CRT Phosphor Characteristics and Applications

Identification	CIE Coordinates x	CIE Coordinates y	Persistence	Fluoresence	Phosphoresence	Decay Time (m/s)	Typical Applications
P-1	0.218	0.712	M	YG	YG	24	Radar and test equipment oscilloscopes
P-2	0.279	0.534	M	YG	YG	35–100	Oscilloscopes
P-3	0.523	0.469	M	YG	YO		
P-4	0.270	0.300	M/S	W	W	25	Monochrome TV
P-5	0.169	0.132	M/S	B	B	25	Photography
P-6	0.338	0.374	S		W		
P-7	0.357	0.537	L	Y	YG		Radar, sonar, and oscilloscopes
			M/S	B	B		
	0.151	0.032	M/S	W	YG	40–60	
P-11	0.139	0.148	L	B	B	25–80	Photography
P-12	0.605	0.394	M/S	O	O		Radar
P-14	0.504	0.443	M	YO	O		Displays where repetition rate is 2 to 4 s after excitation is removed
	0.150	0.093	M/S	B	O		
			M/S	PB	YO		
P-15	0.246	0.439	VS	G	G		TV pickup of photographs by flying spot scanning
P-16	0.175	0.003	VS	BP	BP	0.12	TV pickup of photographs by flying spot scanning
P-17	0.302	0.390	L	Y	BP		Military displays
			VS	G	BP		
			S	YW, BW	Y		
P-18	0.333	0.347	M	W	W		Low-frame-rate TV
P-19	0.572	0.422	L	O	O		Radar
P-20	0.444	0.536	M/S	YG	YG	0.01–2	High-visibility displays
P-21	0.539	0.373	M	RO	RO		Radar
P-22	0.155	0.060	M	B	B	25	Color TV
	0.285	0.600	M	YG	YG	60	
	0.675	0.325	M	OR	OR	0.9	
P-23	0.375	0.390					Interchangeable with P-4
P-24	0.245	0.441	S	G	G		Flying spot scanner tubes
P-25	0.557	0.430	M	O	O		Long-persistence displays up to 10 s
P-26	0.582	0.416	VL	O	O		Radar
P-27	0.674	0.326	M	RO	RO	27	Color TV monitors
P-28	0.370	0.540	L	YG	YG	0.5 s	Radar and sonar
P-29	(P₂ +	P₂₅)	M	W	YW		Aircraft indicators and radar
P-31	0.193	0.420	M/S	G	G	4	Oscilloscopes
P-32			L	PB	YG		Radar
P-33	0.559	0.440	VL	O	O		Radar
P-34	0.235	0.364	VL	BG	YG	40 s	Oscilloscopes and radar
P-35	0.286	0.420	MS	G	B	1	Oscilloscopes
P-36	0.400	0.543	VS	YG	YG	0.25	Flying spot scanning tubes
P-37	0.143	0.208	VS	B	B	0.155	Flying spot scanning tubes
P-38	0.561	0.437	VL	O	O	1040	Integrating phosphor for low-repetition-rate displays and radar
P-39	0.223	0.698	L	YG	YG	150	Integrating phosphor for low-repetition-rate displays and radar
P-40	0.276	0.3117	M	W	YG		Integrating phosphor for low-repetition-rate displays and radar

B = blue, G = green, O = orange, P = purple, W = white, Y = yellow.
VL = very long, 1 s or over.
L = long, 100 ms to 1.0 s.
M = medium, 1 ms to 100 ms.
M/S = medium short, 10 μs to 1.0 ms.
S = short, 1 μs to 10 μs.
VS = very short, less than 1 μs.
 Special Notes:
P-1—High efficiency, resolution, and resistance to burn; lacks low-level or short persistence.
P-2—Decrease in decay with increase in beam current.
P-4—Sulfide version.
P-7—High efficiency and resistance to burn; amber filter required to obtain long persistence.
P-19—Slow refresh rate for flickerless display; low light output; low burn resistance.
P-22—Sulfide blue and green, vanadate for red.
P-25—Desired low-level persistence; high resistance to burn; low light output.
P-26—Slow refresh rate for flickerless display; low light output and burn resistance.

(continued)

P-31—Curve has blue peak at 450 nm; high efficiency, resolution, and resistance to burn; lacks low-level or short persistence.

P-33—Decay decreases with beam current decreases; burns rapidly when used with stationary or slow-moving beam.

P-34—IR stimulatable; Y-phosphor.

P-35—Resists burning compared with P-11.

P-39—Similar to P-1 but with longer decay.

P-40—Similar to P-7 but with longer decay.

P-37—Similar to P-11 but with shorter decay.

P-36—Similar to P-20 but with shorter decay.

Source: After H. R. Luxenberg and R. L. Kuehn, *Display System Engineering*, McGraw-Hill Book Company, New York, 1968.

CRT Flicker Thresholds of the Average Observer

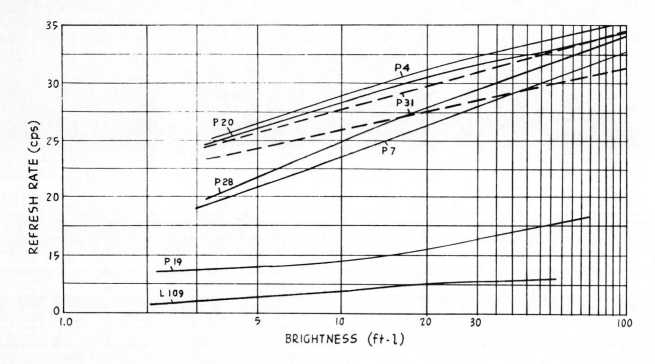

For character displays the pulse rate should be greater than about 30 to 40 Hz so that the characters will not appear to blink. Some flicker may be noticed with average display brightness values unless the repetition rate is maintained at least at 50 Hz.

Flicker will not be noticed in TV at 60 fields per second unless the display brightness exceeds about 180 fL; 50 fields per second is usually acceptable if the display brightness is reduced to 30 fL.

CRT Display Format (Radar and Sonar)

Electronic displays used for detecting, identifying, and tracking target information usually are presented in either rectangular or polar-coordinate form, i.e., an A-scan or PPI (plan position indicator) format. For obvious reasons, the A-scan display face should generally appear as a rectangularly shaped display, and the PPI should appear circular.

A variety of formats are available, as shown in the accompanying illustrations.

Regardless of the type of scope presentation, certain features, such as the following, should be designed according to good human engineering design principles.

DISPLAY SIZE For a single operator involved in a target search task, use a relatively small display (7 to 14 in, 18 to 36 cm) to reduce the area for visual scanning. For tracking targets, use a tube size of 10 to 17 in (25 to 43 cm). If more than one person must work at the CRT station, consider diameters of 24 to 30 in (61 to 76 cm).

DISPLAY CURSORS Cursors help the operator relate target position to actual range and bearing values. They should always be accompanied by indexing scales laid over or around the display face and/or electrically connected to digital panel readouts. Electronic cursors are preferred because they do not produce the parallax problems that mechanical cursors often create.

GRID OVERLAYS Target position can be interpreted much faster if a grid overlay is supplied. The more accurate the reading requirement, however, the more elaborate the grid structure must become. Grids are more confusing than useful on displays that are less than about 14 in (36 cm) in diameter. Even here it may be desirable to create a grid that will not obscure small target traces.

For example, the minimum spacing between range rings on a polar display should be on the order of 1° (36 min visual angle subtended at the eye), or about 0.50 in (1.27 cm) at an average 18-in (46-cm) viewing distance. A mixture of solid and dashed grid lines or range rings helps reduce the potential for obscuring targets and also aids in identifying major range and bearing elements (numbered).

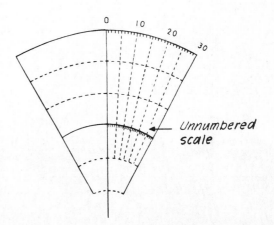

Unnumbered scale

VISUAL DISPLAY INTERFACE
Quantitative Readout Digital Displays

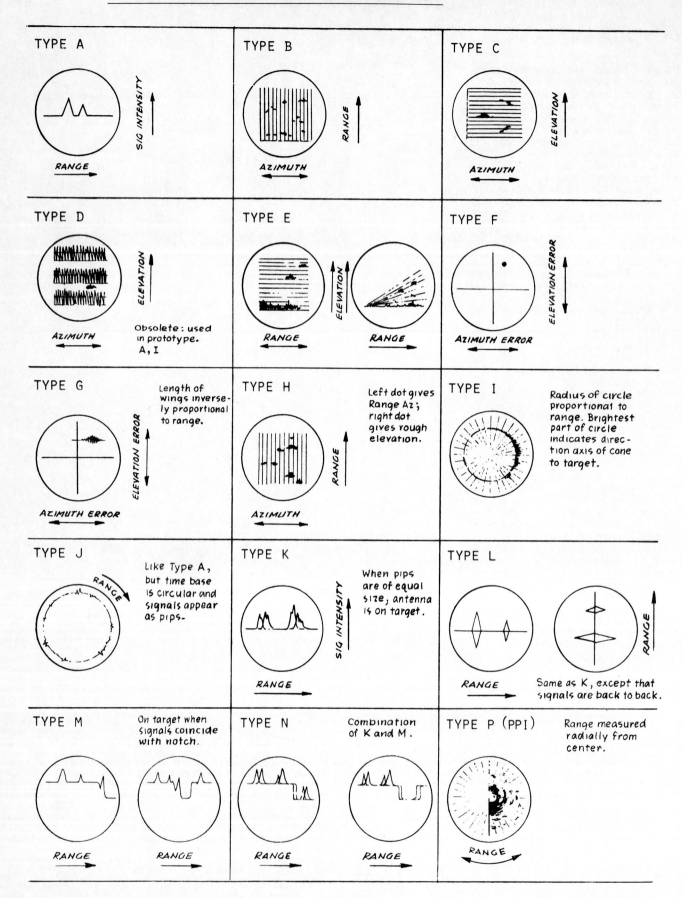

TYPE A

SIG INTENSITY

RANGE

TYPE B

RANGE

AZIMUTH

TYPE C

ELEVATION

AZIMUTH

TYPE D

ELEVATION

AZIMUTH

Obsolete: used in prototype. A, I

TYPE E

ELEVATION

RANGE

ELEVATION

RANGE

TYPE F

ELEVATION ERROR

AZIMUTH ERROR

TYPE G

Length of wings inversely proportional to range.

ELEVATION ERROR

AZIMUTH ERROR

TYPE H

Left dot gives Range Az; right dot gives rough elevation.

RANGE

AZIMUTH

TYPE I

Radius of circle proportional to range. Brightest part of circle indicates direction axis of cone to target.

TYPE J

Like Type A, but time base is circular and signals appear as pips.

RANGE

TYPE K

When pips are of equal size, antenna is on target.

SIG INTENSITY

RANGE

TYPE L

RANGE

RANGE

Same as K, except that signals are back to back.

TYPE M

On target when signals coincide with notch.

RANGE

RANGE

TYPE N

Combination of K and M.

RANGE

RANGE

TYPE P (PPI)

Range measured radially from center.

RANGE

Scope types.

DATA PRESENTATION GUIDELINES

Data should be presented to the operator in a readily usable and readable format. The operator should not be required to transpose, compute, interpolate, or translate to other units or bases. Consider the following:

a. When five or more digits or alphanumerics are displayed (and no "natural" organization exists), group characters into three- or four-character groups, each group separated by one blank character.

b. Identical data should always be displayed in a consistent standardized manner.

c. Alphanumeric data presented in tabular form should be left-hand justified; numeric data right-hand justified by decimal point.

d. Lists should be vertically aligned and left-hand justified, with indentation to indicate subclassifications.

e. Data that require scanning and/or comparison should be in either tabular or graphic form.

f. Minimize use of hyphenation.

g. Individual fields should be labeled (operators should not have to determine field by study of the data content).

h. If a list has to be continued on another page, repeat "last line" as the first line of the succeeding page. Repeat any columnar headings on all succeeding pages. Scrolling should be used to allow an operator to control paging. A minimum of four text lines should be provided in view when display space limits amount of text exposure.

i. Scrolled text movement should be such that the viewer "reads down the page" (as he or she would do with hard copy), i.e., the text is actually moving upward through the screen frame.

j. If the operator is given manual control of page change or scrolling, appropriate switch-label interfaces must be provided. Since page change generally involves shifts from a previous to a succeeding page (new page to the right), two side-by-side keys are appropriate: the left key labeled PREV PG, the right key labeled NEXT PG. For scrolling, an appropriate configuration would be two keys arranged vertically. The upper key would be labeled READ FWD; the lower key, READ BACK.

k. Advisory lists which are being added to on a continuing basis should operate in the following way: the top advisory is removed and the new advisory is added to the bottom (in effect the whole list moves up one). It is helpful to encode the new entry to draw attention to it, i.e., by making it brighter or by placing an asterisk next to the new entry.

MESSAGE INTELLIGIBILITY

To be effective, words have to be selected carefully so that their meaning is perfectly clear to the observer. One should first try to determine the background of the expected user population. The lay person, for example, probably will not understand some words that may be clear to an engineer. Wherever possible, use common words and abbreviations so that everyone will understand the meaning intended.

Word Labels and Instructional Materials

1. Use common terms that originate from typical language usage and/or from standard lists of terms for special fields or groups of users (pilots, military personnel, etc.).

Examples of General versus Special Usage Terminology

General (Preferred)	Special
Speed	Velocity
Rear window	Back light
Rest room	Head
Brightness	Intensity
Adjust	Calibrate
Water analyzer	Ionic chromatograph

2. Use whole words rather than abbreviations wherever space permits. If abbreviations are required, refer to the list of standard abbreviations created by the appropriate "user" group (military, Department of Transportation, etc.). However, check these lists for duplications; it is undesirable to use the same abbreviation for two different functions.

Preferred	Avoid
Range	RNG
Bearing	BNG
Ticket	TKT
Air conditioner	A/C
Transmitter	TM
Tachometer	TCM
Power amplifier	PA

Acronyms may be used sparingly if they have been well established over a long period. They are least desirable for consumer products because of the wide variation in user experience.
3. To identify instruments, use terms that indicate what the instrument measures rather than the name of the device.
OHMS not OHMETER
RPM not SPEEDOMETER
HRS/MIN not TIMER
4. Avoid the use of words that may be interpreted one time as a noun and another time as a verb whenever both may occur at the same workplace.
FIRE = a fire has occurred (warning)
FIRE = to fire a weapon (command)
5. Always use capital letters for labels and short instructions because they can be read at a greater distance than capital and lowercase letters. Capital and lowercase letters should be used only for

extended sentence messages where it is necessary to provide punctuation.
6. Generally, instruction labels should be as brief as possible within the limits of clarity. If an instruction implies several sequential steps, arrange each step on a separate line, preferably with a dot at the beginning of each line.

THIS

· JETTISON CANOPY
· FEET IN STIRRUPS
· RAISE ARMRESTS
· SIT ERECT
· SQUEEZE TRIGGER

NOT THIS

FOR EMERGENCY EJECTION IN FLIGHT JETTISON CANOPY BY PULLING CANOPY JETTISON HANDLE. HOOK HEELS IN FOOT RESTS. RAISE BOTH ARMRESTS TO HORIZONTAL POSITION. ASSUME ERECT POSTURE, ACTUATE EJECTION TRIGGER FOR SEAT EJECTION

The original instruction was:

WALK UP ONE FLOOR
WALK DOWN TWO FLOORS
FOR IMPROVED ELEVATOR
SERVICE

What the instruction really meant was: If you intend to go up only one floor or down two floors, walk instead of taking the elevator because this will relieve the use load and therefore provide better service for everyone.
Following are two ways to simplify the message:

SAVE TIME BY WALKING
IF YOU ARE GOING UP ONE OR
DOWN TWO FLOORS

TO GO UP ONE FLOOR
OR
DOWN TWO FLOORS
PLEASE WALK

7. Avoid the use of words that may have a negative psychological effect; e.g., the word "speed" itself implies traveling at a fast rate and therefore may encourage a motorist to drive fast.

THIS

SLOW TO 10 MPH

SLOW DOWN
WORKMEN IN ROADWAY

SLOW DOWN
ANIMALS IN ROADWAY

DANGER
LETHAL VOLTAGES

SLOW DOWN
ROCKS ON ROAD

NOT THIS

SPEED ZONE AHEAD
10 MPH

SLOW MEN WORKING

CAUTION
DEER CROSSING

DANGER
HIGH VOLTAGE

WATCH FOR
FALLING ROCKS

8. Avoid using the number 1 in combination with letters when encoding such things as airport terminal gates. For example, if you are planning to encode the gates with letter-number combinations, such as K-1, K-2, K-3, etc., do not include the letter I because it is easily confused with the number 1; i.e., I-1 might look like an H from a distance.

9. When special precautionary words are required, select ones that provide a sense of urgency, hazard, or danger. Avoid words that seem to leave it up to the observer to decide whether there is a danger involved, since he or she may not have the necessary experience to make such a judgment.

Preferred Nomenclature	Poor
Warning	Note
Hazard	Precaution
Caution	Slow men working
Danger	Slippery
Watch out for workmen	Slide area
Watch for children	

10. The lettering for all labels should be oriented so that they read from left to right, not around corners, on the side, or up and down.

11. A manufacturer's label must not overshadow the primary display labels and markings. Place the manufacturer's label in an inconspicuous place, in small print, as shown in the illustrations at the right.

12. Control and display labels should be located consistently, either below or above the panel component, especially on the same panel. The decision to locate the label above or below components should be based on the eye position of the observer; i.e., if the panel labels are above the eye reference level, the controls may extend out so far that they obscure a label placed above the control, or the opposite might be true if the control panel is considerably below the eye reference level. Random placement of labels on the same panels makes them difficult to find and associate with the pertinent function.

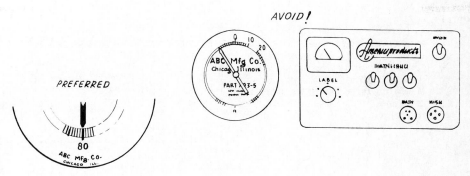

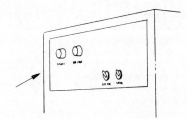

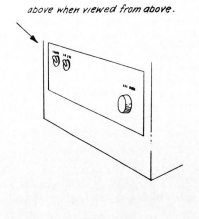

Put labels below controls when viewed from below,

above when viewed from above.

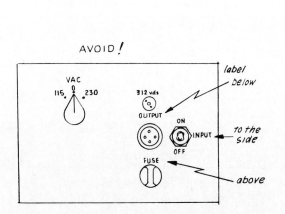

13. Avoid running adjacent labels together. This often happens because control elements were spotted on the panel without considering where the labels would go or how much space they might require.

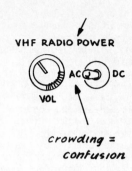

crowding = confusion

14. Utilize size and/or color coding to help the observer differentiate between levels of importance on your panel; i.e., the *system* label should be slightly larger than subsystem labels, and these should be larger than component labels. The examples shown in the accompanying sketches should be used as a general guideline.*

Primary title about half again larger than intermediate title – this half again larger than smallest title.

15. As a general rule, all labels should be made as permanent as possible; i.e., they should be wear- and damage-resistant. If labels are to be read under low light levels, they must be illuminated by means of ambient room lighting, special supplementary flood-lighting, back or edge lighting, and/or luminescent materials or substances. If raised letters are provided (i.e., cast into a panel or package housing), there must be additional coloring to make the letters contrast with the background. Do not assume that a raised letter that is the same color as the basic background will be readable just because it is raised. Engraved or depressed lettering should also be of a color that contrasts sharply with the background. In addition, the engraving should always be filled level with the panel surface so that it will not collect dirt and eventually become unreadable. Avoid glossy inks or paints and plastic or glass dust covers, unless these are specially treated to minimize spectral reflectivity.

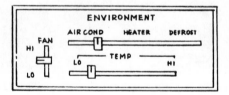

Typical good practice

RAISED

ENGRAVED

*The character height of the smallest label must be sufficient for the maximum viewing conditions expected. Avoid the tendency to settle for some arbitrary policy set by your engineering department for all equipment turned out by the company; this may be completely unsatisfactory.

LABEL SPECIFICATIONS

Starting with the smallest lettering size that will be compatible with the typical average viewing distance, establish specifications for each of the labeling categories illustrated by the accompanying sketch.

Typical Panel Labeling Standards

Label Designation	Letter Size	Location
1. Panel title	18 pt (0.187 in)	Centered; ¼ in from top edge of panel
2. Panel subsection	14 pt (0.156 in)	Centered at top of subsection; ¾ in from top edge of panel
3. Subtitle	12 pt (0.125 in)	¼ in above components or ⅛ in above labels of individual components
4. Toggle switch	10 pt (0.093 in)	¼ in above and below standard switch
5. Single component	12 pt (0.125 in)	¼ in above component
6. Rotary switch positions	10 pt (0.093 in)	¼ in from apex of pointer, line from pointer to label

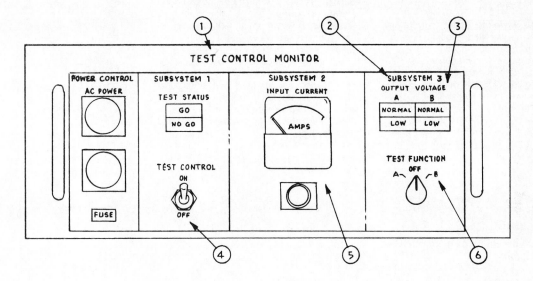

GLARE

Effect of Glare on Performance

Function	Effects
Visibility	Glare of 5 fc located 5° above the line of vision reduces visibility about as much as replacing a 100-fc lighting system with a 1-fc system. An object that can be seen at 20 ft with 5-fc glare at 5° could be seen at 45 ft with no glare. Glare becomes less significant as the brightness of an object is raised, and is greater for a large object seen in low contrast to its background than for a small, high contrast object, such as a printed letter.
Size threshold	Glare greatly increases size threshold, particularly for objects of low contrast. The effect of glare diminishes as contrast and illumination of surrounds are increased, and thus is more easily controlled at high levels of illumination. Threshold size for an object of 5% contrast under 1 fL with a 5-fc glare 5° above line of vision = 30.0 min visual angle; for an object at 100 fL with no glare = 0.63 min.
Contrast threshold	Glare greatly increases the contrast threshold. The degree of increase depends in complex fashion upon the brightness of task background, the glare illumination produced by each glare source at the eye on a plane perpendicular to the line of sight to the task, and the angle of separation between the glare source and the line of sight.
Muscular tension	Glare significantly increases muscular tension—in some cases up to 30%. Although visual tasks may be carried out under severe handicaps with no loss in immediate accuracy, a decrease in efficiency is apparent with a considerable increase in nervous muscular tension. Subjects not compensating for glare effects may show a considerable drop in performance but no increase in muscular tension.
Vision through eyeglasses	The number of persons wearing glasses is sufficiently large and the glare effect sufficiently severe so that reflections from eyeglasses should be considered in placing light sources. When the light source is behind the head, no glare from eyeglasses will exist if the light source is 30° or more above the line of vision, 40° or more below the line of vision, or at an angle of 15° or more with the axis of symmetry of the head.

USE OF COLORS FOR SIGNING AND MARKING SAFETY HAZARDS*

Red should be used to identify:
  Fire protection equipment, fire apparatus, and fire alarm boxes
  Fire blanket boxes
  Fire buckets or pails
  Fire exit signs
  Fire extinguishers
  Industrial fire hydrants
  Fire pumps
  Fire sirens
  Post indicator valves for sprinkler systems
  Cans containing flammable liquids
  Stop bars on hazardous machines
Orange should be used to identify:
  Dangerous parts of machinery
  Energized equipment that could cut, crush, shock, or otherwise injure people
  The inside of a cabinet door to indicate when the door is open, when guards around moving parts are removed, or an electrical circuit box is open, exposing high voltage potential
Yellow should be used to identify:
  Possible physical hazards such as being struck, falling, stumbling, tripping, or being caught between moving and stationary devices. Yellow and black striping around the hazard is recommended
Green should be used to identify:
  Safety equipment
  First-aid equipment
Blue should be used to identify:
  Situations in which equipment should not be started or moved
  Situations in which equipment is being repaired or worked on
  Electrical circuit boxes
Purple should be used to identify:
  Radiation hazards including x-rays, alpha rays, beta rays, gamma rays, neutrons, protons, deuterons, and mesons. Yellow should be used in combination with purple for markers such as tags, labels, signs, and floor marking

Black and white should be used as basic colors for designating traffic and housekeeping markings. Black and white striping and/or checkerboard borders and solid areas may be used for added conspicuousness.

*For additional information, refer to *Safety Color Code for Marking Physical Hazards,* ANSI Z53.1.

RECOMMENDED MUNSELL DESIGNATIONS FOR COLOR CODING VISUAL DISPLAYS

Color-Code Preferences for Four Different Color-Code Schemes (Munsell Color Designations)

Eight-Color Code		Seven-Color Code		Six-Color Code		Five-Color Code	
n	p	n	p	n	p	n	p
1R	999	5R	1008	1R	999	1R	999
9R	892	3YR	890	3YR	890	7YR	884
1Y	946	5Y	1128	9Y	1131	7GY	960
7GY	960	1G	1103	5G	1101	1B	1093
9G	1099	7BG	1095	5B	1087	5P	1007
5B	1087	7PB	1133	9P	1005		
1P	1135	3RP	1003				
3RP	1003						

Note: n = book notation in Munsell Color System. p = Munsell Production Number. R = red. Y = yellow. G = green. B = blue. P = purple. RP = reddish purple, YR = yellowish green, etc.

Source: Adapted from Conover and Craft, *The Use of Color in Coding Displays,* WADC, WPAFB, WADC-TR-55-471, October 1958.

LABELS PLACED ON CONTROLS

Word Displays in Conjunction with Controls

Although perhaps more appropriately discussed under the general topic of "labeling," words used directly on certain types of control devices increase the utility and efficiency of control identification and operation by combining the identifying variables of control position, light, and words. The examples shown in the accompanying illustration are representative of only a few of the applications that would be appropriate. See the section on controls for a more complete discussion of this application of word and light combinations.

Several precautions are in order relative to placing word labels directly on the control device:

The control must not rotate so that the label may become unreadable.

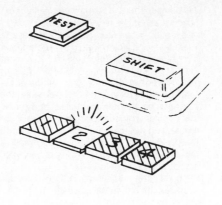

If the control surface area is too small for a fully legible label, place the label adjacent to the control; i.e., do not feel that you have taken care of your obligation just because a label is there—it has to be readable.

Push buttons should extend out from a control panel, as compared with a telelight (i.e., not a control); otherwise, an operator will be unable to tell which similar-looking word-light displays are really controls and which are merely advisory displays.

PICTORIAL SYMBOLS

Pictorial symbols make it possible for persons with different language backgrounds to recognize and understand a single symbol. Unfortunately, it is not always possible to pictorialize all informational conditions since some things would not be recognized by the observer because he or she has never seen the actual object. An example is the choke system used on some vehicles. A symbol that is significant to engineers because they have seen the actual system or device not only is meaningless to average drivers but also may encourage them to believe the symbol implies something entirely different; e.g., the typical hourglass pictorial suggested by the automotive designer could be construed to mean "time" to the lay person. Some principles to consider in deciding whether a pictorial signing method should be used include the following.

1. Pictorial Comprehensibility
If a particular situation includes items that cannot be completely pictorialized, i.e., if some items are easy to pictorialize, but others are not, it is better to stick to a word labeling system; observers will only be confused if they find some pictorials and some word labels on an operator panel.

2. Combined Pictorial and Word Labeling
Whenever there is any doubt as to whether some observers will be able to understand the meaning of a pictorial, use both pictorial and word labels.

3. Size-Distance Compatibility
Some pictorial patterns may be effective only when the viewing distance and lighting conditions are optimum; be sure that a particular pictorial pattern does not lose its identity when it becomes smaller or more distant and/or when the ambient lighting and/or atmospheric conditions are not good.

4. Orientation
Do not place pictorials on components that may be reoriented, i.e., turned or moved so that the pictorial may not appear right side up. Although the observer might figure out what the pictorial is after considerable study, a prime purpose of the single pictorial is to elicit a quick response with a single symbol, as opposed to several words in a label.

5. Sufficient Detail
In creating a pictorial to represent some actual visual element, provide just enough detail to make the symbol recognizable, and no more. Fine detail often cannot be seen under certain distance and lighting conditions and may serve only to distort the impression the symbol creates. On the other hand, be extremely careful not to overstylize pictorial symbols just

to create an artistic rendition. Such stylization often makes all symbols begin to look alike.

6. Isolation
Some type of border should always be used around a pictorial; otherwise, it may blend with background images.

DESIGN OF DIALS AND GAUGES

Dials and gauges are used primarily to provide quantitative information. However, as opposed to a digital display (which provides extremely precise quantitative information), the dial or gauge also provides some additional information in the form of advance warning, rate of change, and/or opportunity to make "cross-dial" extrapolation; this is due to the fact that the pointer position and motion act as an additional qualitative cue as to what is happening.

First, and perhaps foremost, whenever possible avoid the use of dials and/or gauges in which double pointers or scales are used. A single scale and pointer format is best because it provides the least reading error.

As shown in the accompanying sketches, both round (or half-round) and linear or rectangular formats are possible. It has not been shown that either format results in any difference in readout effectiveness.

However, the accompanying illustrations comparing the round and rectangular gauge formats provide some interesting insights into the possible value of the rectangular format over the round format:

When several gauges are placed side by side, it is obvious that the relative positions of the pointers are more readily assessed.

The direction of pointer increase and/or the value level is always clear in the rectangular format, whereas in the round dial, it is more difficult to tell by the position of the pointer just where it is, depending on which quadrant of the dial the pointer is in; i.e., when the pointer is on the left or top half of the dial, the direction of pointer movement for increase and decrease is "natural," but when the pointer is in the right or bottom half of the dial, the increase and decrease direction of motion is reversed.

In the case of an automobile speedometer, for example, the horizontal scale can be placed higher on the instrument panel (that is, the entire scale can be placed higher); this makes it easier for the driver to glance at the instrument without taking his or her eyes too far from the road.

Often, several rectangular gauges can be placed closer together than round dials. This, of course, depends on the particular instrument package shape behind the panel.

When several round dials have to be clustered for ease in check readings (i.e., when an operator must glance at them quickly to ascertain that things are functioning properly), the instruments can be oriented so that all the pointers are aligned alike when the instruments are indicating normal operation. This makes a single pointer misalignment easy to see.

Vertical arrays should be oriented so that normal pointer positions are vertical; horizontal arrays should be oriented so that normal pointer positions are horizontal.

Dial Size

The size of a dial or rectangular gauge normally should be determined by the number of scale graduation marks required (which is a matter of the inherent precision required). Instrument manufacturers have tended to standardize their instrument package sizes for the sake of production convenience and therefore have sometimes put more markings on the instrument face than can be accurately read. The spacing of scale graduation marks must be great enough so that the observer can discriminate between one mark and another, and also see the relation between the marks and the pointer, without taking an inordinate amount of time peering at the instrument.

In addition, the size of the dial depends on how deeply inset the face is below the bezel. When the manufacturer's case design includes a deeply inset dial face, the bezel tends to limit the viewing angle, with the result that (depending on the viewing angle) some of the numbers and/or scale marks may not be visible. Some designers have created poor formats to overcome this problem, such as placing the numbers "inside" the scale to make sure that they are not obscured. As will be noted later, this format is counter to good pointer-scale relationships and results in a pointer that partially covers up the scale numbers.

Dial Size in Relation to Required Number of Scale Marks

Too many scale marks, too close together, make dial reading difficult and error-prone. In fact, it is impractical to have a dial that is so small that the inner annulus of the scale is less than 1 in (2.5 cm) in diameter. Use the accompanying table as a guide in determining the size of dial necessary to accommodate the required number of scale marks. For example, if the viewing distance is approximately 12 ft (3.7m) and if 50 scale graduation marks are required, the inner annulus of the scale must be at least 5 in (13 cm) in diameter.

Minimum Diameter of Inner Annulus at Various Viewing Distances*

No. of Scale Marks	Viewing Distance				
	20 in	3 ft	6 ft	12 ft	20 ft
50		1.3 in	2.6 in	5.0 in	9.0 in
100	1.4 in	2.6 in	5.0 in	10.0 in	17.0 in
150	2.0 in	3.9 in	8.0 in	15.0 in	26.0 in
200	2.9 in	5.0 in	1.0 in	21.0 in	34.0 in
250	3.5 in	6.4 in	13.0 in	26.0 in	43.0 in
300	4.0 in	7.7 in	15.0 in	31.0 in	51.0 in
350	5.0 in	9.0 in	18.0 in	36.0 in	60.0 in

A common design error occurs when the designer believes that a specified level of accuracy must be provided and therefore crowds the scale marking in order to get a certain number of marks within the constraints of the dial-face diameter, as shown in the accompanying illustration. The result is that the observer cannot differentiate between the marks and not only makes errors in reading but also spends more time reading; thus the basic objective is lost.

The designer should either reconsider the need for the level of accuracy thought to be required (i.e., the instrument itself may not provide the level of accuracy it was assumed to) or use fewer marks, since the observer can probably interpolate the pointer position between fewer, properly spaced marks more accurately and reliably than he or she would be able to do with the poorly marked dial.

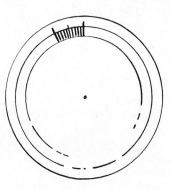

POOR

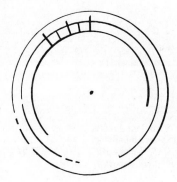

PREFERRED

Scale Marking

The following are some typical scale-marking design errors to be avoided:

1. Dots instead of lines
2. Thick marks
3. Marks joined by a heavy base line
4. Long marks spaced close together
5. Irregularly spaced marks about an arc, which cause the observer to shift his or her spatial reference
6. Uneven spacing, which leaves the observer in doubt as to the value of each mark
7. Significant spacing variation between the ends of the marks, which, because of the tight radius of the dial, makes it hard for the eye to follow along the scale-pointer axis

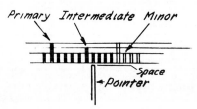

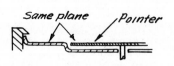

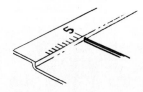

Pointer-Scale Relationships

For ease of reading, the scale-pointer relationship should be as shown in the accompanying sketch. The point tip should be the same width as the scale mark, and it should fall just short of the mark.

Placing the pointer just inside the scale annulus ensures that it will never be confused with a mark. However, the plane of the scale and the pointer also should be the same; otherwise, the observer may misread the scale-point association because he or she was looking at the instrument from an angle (parallax). The accompanying sketches illustrate the proper arrangement for pointer and scale planes.

Finally, the numbers associated with the scale should be outside the scale ring so that the pointer will not obscure the number.

Instrument Scale-Marking Guidelines

Experiments by military psychologists produced the scale-marking specifications shown below; scales were read with fewer errors and faster response when these dimensional characteristics were present. The 28-in (71-cm) viewing distance is typical of the aircraft cockpit situation. For other viewing distances, the proportions should be maintained, even though values are increased or decreased; i.e., multiply each dimension times the viewing distance, or

$$\text{Dimension at 28 in} \times \frac{x \text{ in}}{28}$$

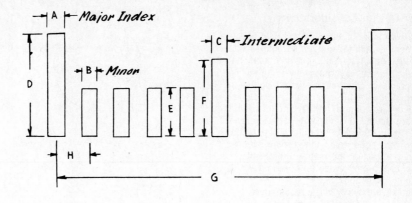

Scale Mark Dimensions for 28-in (71-cm) Viewing Distance

Dimension	Black on White	White on Black
A width	0.035 in (0.069mm)	0.125 in (0.032 mm)
B width	0.025 in (0.054mm)	0.125 in (0.032 mm)
C width	0.030 in (0.076mm)	0.125 in (0.032 mm)
D length	0.22 in (0.56mm)	0.22 in (0.56 mm)
E length	0.10 in (0.25mm)	0.10 in (0.25 mm)
F length	0.16 in (0.41mm)	0.16 in (0.41 mm)

Note: Although the major, intermediate, and minor index mark widths are shown to vary for black on white markings, it is suggested that all mark widths be equal for most applications so that a pointer tip can be used that is the same width as all the marks. Use the minor index width to define the pointer tip.

Direction of Pointer Movement

Scale numbering should reflect the expected direction of pointer movement; i.e., pointers should move to the right, up, or clockwise to indicate a value increase.

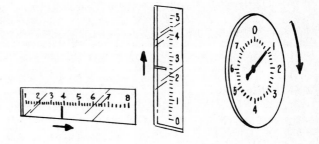

Scale Interval

For most people, scale intervals of 1 through 10 (or multiples thereof) are easiest to interpret without error. The next easiest are intervals of 2, 4, 6, 8, and 10. However, intervals such as 3, 6, and 9 (or 4, 8, and 12) can cause confusion.

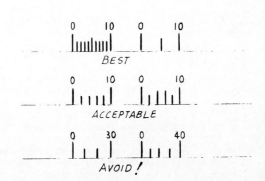

Other Design Considerations

Make sure that numerals are oriented upright rather than slanted and turned on their side. The exception to this rule is the situation in which the dial may rotate, in which case it is desirable to orient each numeral so that its base is toward the center of the dial, which makes the numeral appear upright when it rotates to an indexing position, preferably at the top of the dial. Alternative orientations may be necessary if the display exposes only one-half of the dial and the display is oriented to emphasize an up-and-down direction of motion.

If a scale covers only part of a circumference, attempt to balance the number positions so that they present a symmetrical appearance.

Consider the use of patterned or colored range strips when the primary readout is to determine when a system is performing within specified ranges, i.e., when it is less important to read specific values.

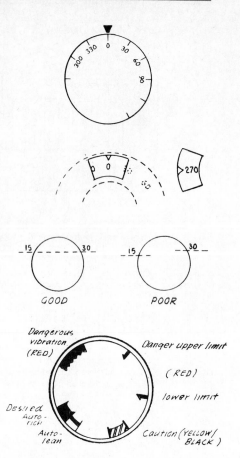

Although it is always desirable to optimize the size and shape of numerals used on dial displays, circumstances may dictate a smaller instrument face. It is better to use smaller numerals and thus avoid the clutter and illegibility that result when characters of a so-called optimum size are crowded together.

Whenever the dial scale is of finite length, be sure to leave a sufficient break between the beginning and end of the scale so that observers do not mistakenly assume that the scale continues and/or that they should continue reading around the scale, but with some special extrapolation.

When several dial or rectangular, scaled instruments or gauges are mounted within a single work station or on the same instrument panel, make sure that the scale breakdown is alike on all the instruments; i.e., do not mix two different scale breakdowns, as shown in the accompanying illustration.

Instrument faces for use when night or dark adaptation is not critical are more easily seen if the markings are black on a light-colored face. Even though the instrument may have to be illuminated at night (e.g., an automobile speedometer), the lighter background can still be dimly lit and provide a more readable display. In the case of the automobile speedometer, the display is also easier to see when sunlight reflects on the cover. For blackout conditions, however (airplane cockpits, etc.), light characters and marks on a black background are mandatory to help the observer maintain proper dark adaptation.

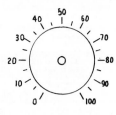

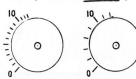

Design of Pointers for Dial and Gauge Instruments

For maximum reading accuracy, a pointer should be of equal width throughout so that the parallel sides visually project to match the parallel-sided scale marks to which the pointer must be associated.

The width of the pointer should be exactly the same width as the scale marks. That is why, in a previous discussion of scale marking, it was noted that all scale marks should be the same width.

In many applications, however, the thin pointer is not conspicuous enough, especially when the user has to depend on a brief, quick glance to see how a particular system is performing. Thus, it is necessary to increase the width of the pointer, i.e., to make it bolder and more obvious during a quick scanning look. In such cases, the pointer should be designed as shown in the accompanying sketch. Note, however, that although the main stem of the pointer is enlarged, it is still tapered at the tip so that the width of the tip still matches the scale-mark width.

Generally speaking, if a pointer has a "tail" on it, the observer can judge slight movements of the pointer better than if the pointer is tailless. The tail should not be more than one-third the length of the pointing segment, however; otherwise, there may be confusion as to which end is the pointing end.

Above all, avoid artistic pointer designs.

Pointer illumination is often difficult. A translucent dial face provides a good background for silhouetting a black pointer. Another useful technique, shown in the accompanying sketch, consists of a plastic pointer ring inside a scale ring. Both can be either back-or edge-lighted with considerable success.

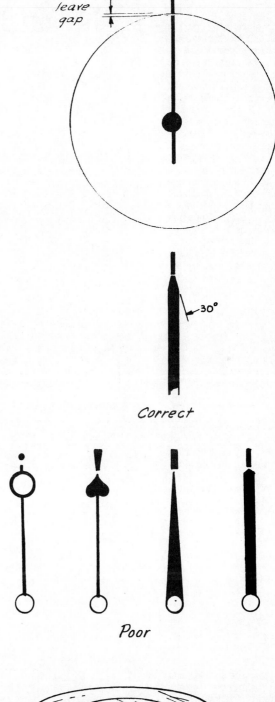

leave gap

30°

Correct

Poor

Example of a Poorly Designed Aircraft Instrument versus the Correct Format

The accompanying illustration of a doppler navigation display shows a typical layout created by an instrument manufacturer.

Both scale breakdowns should have been similar.

Numerals should have been positioned outside of the scale marks so that the respective pointer will not cover marks or numerals.

The two major human engineering errors in this format are typically the result of the designer's attempt to create symmetry (i.e., positional balance of the numbers for the outside scale). The accompanying sketch illustrates how the instrument face should have been laid out. Although symmetry is desirable in certain cases, it is irrelevant here. It is more important to maintain a similar scaling factor between the two scales. Equally important is the principle of keeping pointer-scale and number relationships separated so that the pointers do not obscure any scale marks or numbers, or at least those which must be visually associated. Note that, in the second sketch, the smaller, or inner, scale does not contain the lesser scale marks because they are too close together. Pilots can estimate these as accurately as they can actually "read" the marking-pointer relationship, and they can do it faster. "P" and "S" refer to port and starboard degree readings.

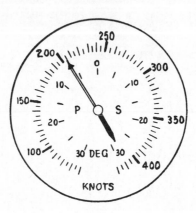

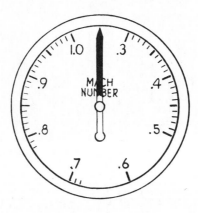

The typical Mach Number display format shown in the accompanying illustration ignores several important human engineering principles. The scale break is at the top of the display rather than at the bottom; i.e., the zero position should be at the bottom so that the initial direction of pointer movement (e.g., upward) is naturally associated with an increase in speed.

Correct layout

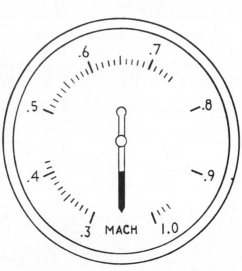

In addition, the numerals, scale marks, and pointer are arranged so that the pointer covers up the marks and numerals. The inner scale-mark annulus should be "even" so that the pointer tip falls short of any of the marks. The numerals should be located on the outside of the scale markings.

Color as an Indicator of Operational Condition

Red: Red should be used to alert an operator to the fact that some situation makes the system inoperative, e.g., an error, a failure, or a no-go or life-endangering condition.

Yellow: Yellow should alert an operator to a situation in which caution, recheck, or delay is necessary, i.e., a condition that, if not attended to, could lead to a dangerous situation.

Green: Green should indicate that equipment is operating satisfactorily or that one can proceed. It can be used to indicate the successful completion of steps within a process, thereby establishing a basis for continuing with the next step.

White: White should indicate such items as status, alternative functions and selection modes, a test in progress, or any similar items that imply neither success nor failure of system conditions.

Blue: Preferential use of blue should generally be avoided because it has no standard meaning, except when it has been assigned a special significance within a given operating system; e.g., a blue reference light has been used by some automobile designers to indicate that headlights are in the high-beam condition.

A flashing red light should be used when an extreme danger is present; i.e., the flashing light indicates a situation of more importance than a continuous red light. A flash rate of between three and five flashes per second, with approximately equal on and off times, is recommended.

The flash coding technique is also suggested for other colored lights when it is desirable to increase the conspicuousness of the signal. Typical examples include a flashing amber traffic signal in advance of a regular stop-lighted intersection and an alerting signal on a communications status board.

For additional suggestions regarding the use of color coding for light indicators, refer to U.S. Army MIL-HDBK-759.

Specific Functional Requirements for Warning, Caution, and Advisory Light Indicators

In military equipment installations, it is required that master warning and/or caution light indicators be placed directly in front of operators to ensure that they will not miss seeing the light when it illuminates. Illumination of such a light indicates to operators that they should immediately look for specific indicators that will tell them what is wrong. These latter indicators are typically located where they can be functionally grouped. Some of the types of warning, caution, and advisory functions that are identifiable by means of annunciator light displays are listed below:

Warning Indicators (Red Light)
Bailout
Cabin pressure failure
Canopy unlocked
Emergency fuel failure
Engine overheated
Engine on fire
Fuel system failure
Landing gear lower failure

Caution Indicators (Yellow Light)
Aileron trim failure
Anti-skid-off
Automatic pilot failure
Constant speed drive failure

Defroster failure
Door open
Engine icing
Generator #- Inoperative
Oil pressure low

Note: Although these examples relate to aviation, the principle of devising separate master and functional indicators to warn, caution, or advise an operator through the use of colored lights has obvious application to a host of situations, including commercial systems. The concept is especially applicable to highway vehicles, communication centers, and such domestic products as farm or construction machines, stoves, and washers and dryers.

Advisory Indicators (White, Green, or Blue Light)
Alternate current
De-ice
Direct current
External power on
Marker beacon
Radar on
Trimmed for takeoff

FLASHING LIGHTS

Flashing lights are recommended wherever it can be anticipated that an observer needs reminding that a situation exists or is about to occur; in certain cases this need arises when there is a high probability that the observer may be visually occupied and needs the benefit of a more conspicuous visual signal. The typical applications and conventions are as follows:

A preparatory flashing amber light in advance of a traffic signal.

An anticollision flashing light on aircraft because of the special alerting qualities of a flashing (and/or rotating) light. Although the basic anticollision light is typically red, high-intensity white flashing lights have been shown to be conspicuous at greater distances.

Flashing lights (both red and blue) for police and emergency vehicle identification and alerting.

A coded flashing light to aid identification of an airfield.

Note: The main benefit of the flashing light is that it is more conspicuous. On the other hand, too many flashing lights easily destroy the conspicuousness of any one signal. The same is true of color. Red was at one time the most effective color to use for taillights because they were easily identified among the typical white street lights. On a crowded freeway, however, there are so many red taillights that the motorist often has difficulty knowing what is going on ahead.

SIGNAL BRIGHTNESS

The desired brightness level of any signal light depends on the ambient lighting conditions, the brightness of other lights in the field of view, and the need to identify a change in situation (i.e., a given signal may be made to glow brighter when a functional change has occurred).

Often it is helpful to increase the conspicuousness of a conditional change by combining a brighter light with flashing. Care should always be exercised, however, not to blind an observer at a critical moment by increasing the brightness of a light. A typical case in point is the change in brightness of the taillight of a car when the brakes are applied. Although the objective is to make the change in brightness sufficient to let the observer in the rear know for sure that the brake has been applied, the observer may be blinded by an extremely bright light and make precarious evasive maneuvers that are not warranted.

LIGHT SIGNAL POSITION

To be effective, of course, a signal or pilot light has to be in a spot where the observer is most likely to be looking. Other positional considerations are also important, such as the following:

A power pilot light should be located in the place where the observer first looks at the equipment (typically on the upper part of the equipment).

A high-beam indicator light should be located at the top of the vehicle instrument panel near the observer's typical line of sight as he or she watches the road. The same applies to turn indicator lights.

Perimeter-defining lights on vehicles should be as near to the edges of the vehicle envelope as is practical; otherwise, the observer may come too close.

Traffic signal lights should be positioned high enough so that they can be seen over the tops of other vehicles, but not so high that when the driver stops, he or she cannot see the signal because of the limits of the vehicle's upper windshield frame.

Warning lights (and/or lighted indicator lights) should be located within about 15° of the principal line of sight of an operator in order to make sure that the operator will notice a warning while concentrating on something else (e.g., watching the road ahead, in the case of a motorist, or watching a CRT display, in the case of a sonar or radar operator).

Stop lights used on the rear of automobiles should be placed at a height above the road approximating the eye level of the driver of the car behind (or slightly above this level). This has been shown to reduce the incidence of rear-end collisions.

COMBINATIONS OF PILOT LIGHTS, ANNUNCIATORS, AND LIGHTED PUSH BUTTONS

It is important to an operator to be able to tell quickly which lighted indicator is a pilot light, an annunciator, or a telelight and which is a lighted push button. Avoid the temptation to select hardware components that all look alike; e.g., use a round fixture for a pilot light, a rectangular fixture for an annunciator or telelight, and a square fixture for a push button.

The accompanying illustration shows how much easier it is to tell the difference between components when each has a different shape.

It is also desirable to select push-button hardware in which the button extends about ⅛ in (0.32 cm) out from the panel and annunciator hardware that is more or less flush with the panel. The best pilot light is one that also extends from the panel so that an operator can tell from an angle when the light is on.

Some of the new flush control concepts that require the operator merely to touch an outlined section of a panel do not comply with the above suggestion. There are no studies to indicate whether operators are confused by such concepts. However, one can be certain of no confusion if the above suggestions are followed.

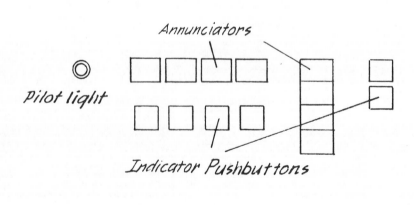

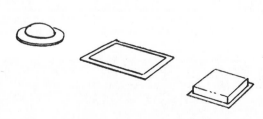

Manual Control Interface

COMPUTER INTERACTIVE DISPLAY-CONTROL

Selection of the appropriate type of control is the first step in insuring that an operator can execute the commands necessary to produce the information he or she wants. The appropriate control can be manipulated effectively and insures the maximum speed and accuracy potentially inherent within the user-machine system.

Each type of control demands some type of manual contact by the operator and therefore must possess characteristic size, shape, and motion features that are compatible with human anatomic capabilities and limitations. The size, shape, and mobility of a person's hands have direct bearing on the ease with which an operator can touch, grasp, hold, and manipulate a control device.

The location and position of controllers are also important in determining the ease and effectiveness with which an operator utilizes a control. Several factors to consider are whether "handedness" may influence controller operation, whether speed and accuracy of movement is affected, whether range of movement is critical, and finally, whether there may be an important spatial relationship between the controller motions or location and motions or location of displayed elements.

MANUAL CONTROL GUIDELINES

General Requirements

Computer program and display-control interfaces should be designed to provide an accurate and rapid interaction between the operator and basic hardware and software. To accomplish this, it is critical to balance the needs of an operator who wants to interrogate, retrieve, store, edit, and compose data in as natural a manner as possible with the capabilities and limitations of the technological state of the art. Avoid creating an operator

task which is machine-oriented and which requires the operator to think differently than he or she would if the task were to be performed using previously learned methods with standard materials, documents, and tools. Obviously one cannot do this in all cases, but this should be one of the design objectives.

Some key operator-related considerations are:

a. Provide continuous feedback to the operator.
b. Provide an easy means of correcting erroneous entries made by the operator.
c. Present informative "error messages"; tell the operator what is wrong and what to do.
d. Hopefully provide error warning before an ENTER action has been taken by the operator.
e. Provide internal software checks to minimize operator errors.
f. Provide both mandatory and permissive PROMPTS to insure correct operator response.
g. Consider use of auditory signals to complement visual indications.
h. Present words and messages in "plain" English, with as few abbreviations as practicable. Use *standardized abbreviations* when required.
i. Use pictorial graphics to speed up communication and to minimize long printed descriptions or messages. However, be sure graphics are recognizable and meaningful to the user (background training or experience).
j. Minimize data entry keying by abbreviating lengthy inputs—avoid ambiguity.
k. Use abbreviations of consistent, fixed length when practicable. If an isolated abbreviation is not clear within the established three-to-four letter format, add one letter to clarify.
l. Avoid abbreviations that may have two meanings.
m. Alarm messages should be distinctively coded so they are readily differentiated from other messages.

n. Alarms should require operator acknowledgment.
o. Command language should reflect the operator's point of view, not the programmer's.
p. Provide a computer-generated command index to help operator compose correct commands.
q. A dedicated display space should be provided in which commands are posted as keyed in. Preferred position is directly above a keyboard.
r. Simplify displays in terms of detail—no more detail than the operator actually needs or can use. When different levels of detail are used, make format conventions similar (e.g., line type, color, shape, etc.).
s. Use computer to simplify operator's task, e.g., acquire readily available information rather than require it to be keyed in; relieve the operator of the burden of routine calculation; automate cross-filing and updating so the operator does not have to enter data more than once.
t. Provide left and/or right justification automatically.
u. Minimize the need to memorize mnemonic codes.
v. Highlight items the operator is manipulating.
w. Use intelligible labels to identify data groups or messages; locate labels in consistent fashion; highlight; place label at top of data group; labels should be "white" when other information is colored.
x. Computer response to discrete control inputs by the operator should appear essentially instantaneously—0.1 second or less for alphanumeric entry; 20 ms or so for continous control function of a tracking nature (slow system responses should be provided with an acknowledgment that a response was initiated).
y. Fixed function (dedicated) keys should be used for time-critical, error-critical, or frequently used control inputs. In simple systems, they are always preferred over

multi-function keys of command language input routines. Activated keys should display some indication of that fact—brighter illumination, for example. A graphic of dedicated keys should be displayed on the bottom portion of the CRT being activated, with the activated key showing brighter than other symbols. If the array can be changed to another mode, the remote display should reflect the change.

Manual, Computer-Input Devices

Design or selection of the appropriate control for data input and/or computer inquiry should be based on the specific type of task. In some cases the tasks may require more than one type of controller. Current industry practice typically employs one or more of the following, for the indicated reasons.

A. TYPEWRITER KEYBOARD—Used for typical, rapid typing of letters and extended message commands. A standard "QWERTY" key arrangement makes such an entry device appropriate for tasks such as word processing.

B. NUMERIC KEYPAD—Used for typical numeric mathematic or calculator-type input. Although there is some question whether the order and arrangement of numbers is optimum (i.e., some studies indicate the telephone system with number one starting at the top left is faster and more error-free), general practice seems to favor the calculator format (i.e., number one starting at the lower left of the button array); the calculator format should include an ENTER key that is larger and to the right of the basic numeric key array.

C. DEDICATED-KEY–PUSHBUTTON ARRAYS—Such arrays are often placed on the same package with QWERTY and keypad elements, either directly above the typewriter array or to the left of the keypad array. In some instances, dedicated keys are designated within the typewriter array at either end or within the upper key row as shift-key functions. Dedicated key operations are not typically sequentially or operationally operated (as in typing or calculating), but rather operated one at a time. They do not require special skills. On the other hand, improper placement among the typing set complicates search for the right key and, in many instances, introduces typing error. It is recommended that when a combination keyboard is required, the following considerations be noted:

a. Primary dedicated key functions should be placed above the standard QWERTY array (with a noticeable space between the arrays). No dedicated keys should be intermixed in the QWERTY array.

b. A numeric keypad should be located to the right of the QWERTY array (with noticable space between the arrays). Certain dedicated keys may be added to the left side of this keypad (e.g., cursor-controls), but these should be visually isolated from the primary keypad keys by varying the color.

D. OTHER, CONTINUOUS-CONTROL DEVICES—A variety of control devices have been successfully used for manipulation of continuous functions such as movement of cursors, manipulation of display graphic position, and orientation on the display screen. Among the more effective devices are the omni-directional joystick, rolling ball, and the "mouse." Perhaps the most effective of these is the mouse since this device allows the operator to maintain a constant grasp that is generally more natural and less subject to inadvertent input motions caused by variations in the user's finger-hand-arm actions. This device also provides an oppurtunity to add discrete switches to the device so that they can be operated with the same hand while the device is being moved about (generally without disturbing the accuracy of the movement or positioning of the device). Switch operation on joysticks typically requires the operator to change hand position slightly, which can disturb the joystick position. Two other common control input devices or techniques are the light pen and the finger-contact screen system. Combinations of any of the above are also possible for special applications. Important considerations in deciding which device to use are these:

a. Joysticks tend to be "preferred hand" dependent, are subject to inadvertent movement input, and occupy a specified fixed space on the workplace surface. Added switching functions are required to allow the operator to change display movement rate, unless switching is given to another hand or foot.

b. Rolling balls also require a separate switch to "hook" a displayed cursor position, and they require considerable dedicated workplace surface. They do not provide a clear kinesthetic feedback as to the position of the controller or the direction in which the controller is being rotated (practice generally overcomes the latter difficulty). This device does provide for very smooth continuous control of display movement rate (little disturbance due to anatomic changes in finger-hand-arm positioning).

c. The mouse-type device provides a very natural and comfortable interface, does not take up dedicated workplace space, is generally operable equally well by either hand, and is less subject to inadvertent inputs due to change in hand posi-

tion on the controller. If properly shaped, the mouse can also contain additional switching functions within the package itself, although care must be taken to use the right switch types, positions, and motions to preclude inadvertent movement being input into the basic device.

d. Light pens are fairly "preferred hand" dependent, but are natural in terms of use and direct relationship to material displayed on the screen. Although no dedicated workplace space is required, some specific place should be provided to stow the pen when it is not in use. The pen cord also presents a minor interference problem in some workplace configurations.

e. Direct finger-contact concepts are perhaps the most natural form of control, but they present a problem in terms of contact point size. Action switching must be accomplished by the other hand or by other means in most cases. Since both this technique and the pen controller involve touching the display screen, some method is required to minimize damage to the display screen or disturbance of CRT normal operating characteristics.

JOY-BALL CONTROLLER

Although slightly less desirable than the pencil joy stick from the point of view of operator accuracy, the joy-ball controller is sometimes preferred in order to eliminate the interference problem occasionally associated with the protruding joy stick.

A separate switch should be provided to disengage the joy ball when it is not being used.

Joy balls are not self-centering and therefore have the disadvantage that one can easily leave the controller in a position in which the display hook or cursor is completely out of view. When joy balls are used, it is desirable to design the control system so that the displayed elements will always remain in view on the CRT.

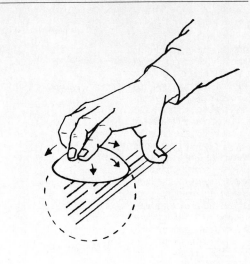

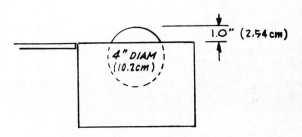

"Mouse" Controller

The "mouse" positioning controller is a fairly recent development designed to replace the other types of coordinate position designation and cursor positioning. When compared with other devices, the mouse is easier to use, faster, and more accurate.

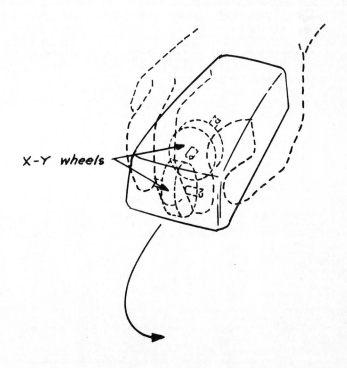

OTHER CONTROLS

A common type of panel-mounted control is the pull-type knob. Although one might expect to push the control forward for ON, GO, or OPER-ATE conditions, long-standing convention has created an opposite expectancy; i.e., the control is pulled out for an operational mode and is pushed in to cause the operation to cease.

Recently, slide-type controls have been introduced for a variety of control operations (including control of vehicular functions, such as the environmental system, and control of radios, e.g., station selection and volume control). When slide controllers are used (not necessarily recommended as the preferred type of controller), the direction of control movement should be as shown in the accompanying illustrations.

ON/OFF function switches, regardless of their particular configuration (toggles, rockers, etc.), should operate as shown in the accompanying illustrations; i.e., movement of the control upward or to the right or depression of the top or right portion of the rocker switch should cause an ON condition.

Rotary-motion switches (including both the detented, discrete-positioning and the continuously variable potentiometer types) should move clockwise for an increase in "value." However, this is true only when the control knob is operated only by the right hand and/or is located on a panel facing the operator.

When such a control is positioned for left-hand use only, the direction of motion should be as shown in the accompanying illustration because the operator shifts from a clockwise stereotyped expectancy to a spatial expectancy relationship; i.e., the operator rotates the control surface forward for an increase in functional value.

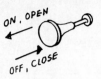

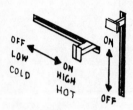

INCREASE

Left hand only

Control position–direction-of-motion relations may vary depending on the location of the control with respect to the operator.

Although, as a general rule, one can translate control motions from one operating plane to another merely by assuming that the operator "faces" the control panel, this does not carry through to all conditions (as shown in the accompanying illustration). Typically, controls alongside an operator are not "faced" until the control position moves aft of the operator's chest. Once aft of this point, the operator typically turns to face the control.

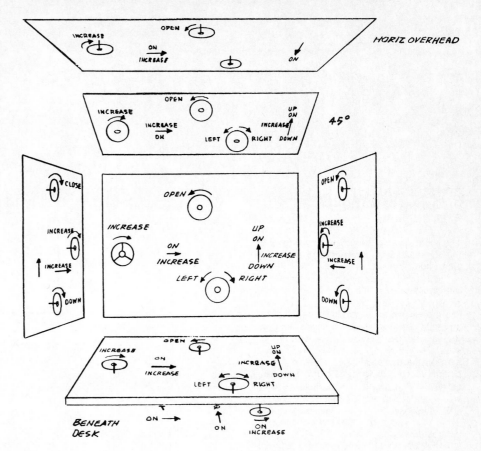

Rotary controls for discrete positioning should use the well-established principle of "a moving pointer versus a fixed scale," as shown in the first illustration. The second illustration shows that when values are imprinted on the knob skirt, they progress in the opposite direction to that in which one normally expects to read them (i.e., the principle of "a fixed index versus a moving scale"). The latter configuration leads to frequent positioning errors.

PREFERRED

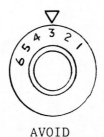

AVOID

CONTROL AND DISPLAY DIRECTION-OF-MOTION RELATIONSHIPS

The most reliable expectancy occurs when the right-hand arrangements illustrated by the two upper sketches are used.

A weaker expectancy occurs when the arrangements for left-hand operation shown in the lower sketches are used.

These expectancies occur because of the apparent mechanical relationships that the operator observes; i.e., the operator, purposely or not, observes the directional movement of the control perimeter and the scale pointer as moving together. In the upper vertical display, the pointer must be adjacent to the side-mounted control. In the lower vertical scale, the pointer is moved to the other side so that it, too, is adjacent to the left side-mounted control.

The lower vertical scale-control arrangements are less desirable because the visual expectancy conflicts with the normal "blind" control movement expectancy noted earlier. Thus, if by chance the operator did not actually look at the scale (i.e., adjusted the controls blind), he or she would probably turn the knob in a clockwise direction for "increase."

Note: Because of the circular geometry of both scale and control knob, left-hand expectancies are the same as right-hand arrangement.

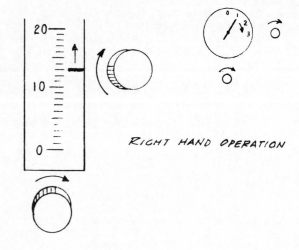

RIGHT HAND OPERATION

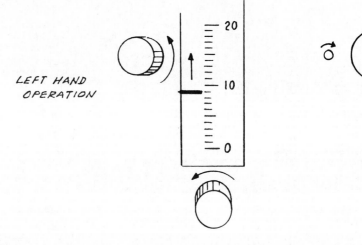

LEFT HAND OPERATION

PUSH BUTTONS

Grouping Push Buttons

As indicated by these guidelines, adequate button separation is required to avoid inadvertent operation of adjacent switches.

Minimum Button Separation

Arrangement	A	B	C	D	E	F
Vertical plane:						
No gloves	0.75 in (1.9 cm)	0.625 in (1.6 cm)	0.1875 in (0.476 cm)	0.25 in (0.64 cm)	1.25 in (3.17 cm)	0.75 in (1.9 cm)
With gloves	1.75 in (9.45 cm)	1.625 in (4.12 cm)	0.375 in (0.95 cm)	1.25 in (3.17 cm)	2.25 in (5.72 cm)	1.75 in (4.45 cm)
Horizontal plane:						
No gloves	1.0 in (2.54 cm)	0.875 in (2.22 cm)	0.4375 in (1.11 cm)	0.50 in (1.27 cm)	1.50 in (3.81 cm)	1.0 in (2.54 cm)
With gloves	2.0 in (5.1 cm)	1.875 in (4.76 cm)	0.375 in (0.95 cm)	1.25 in (3.17 cm)	2.25 in (5.72 cm)	2.0 in (5.1 cm)
Under severe vibration or oscillation	3.0 in (7.6 cm)			3.0 in (7.6 cm)		3.0 in (7.6 cm)
For blind selection	6.0 in (15 cm) apart in front of operator; 12 in (31 cm) apart when buttons are located in the peripheral areas.					

Note: The above guidelines apply to any shape of button.

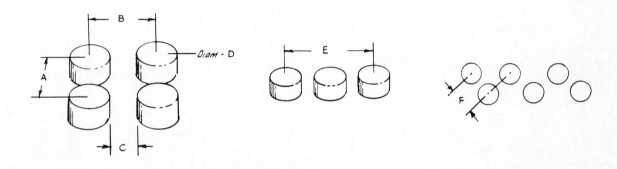

Push-Button Shape

From a biomechanical point of view, it makes no difference whether one selects a round, square, or rectangular push button. However, the square or rectangular button provides more useful area for labeling, and obviously a rectangular button laid on its side provides more label space than one that is standing on end.

Shape can also be used as a cue for determining the functional significance of a particular push button. This is especially important when advisory indicators are used in combination with push buttons. A round push button is seldom mistaken for an advisory indicator.

Off-the-shelf push-button configurations such as those shown in the accompany illustration are popular for many applications. The separator bar between buttons is also helpful in preventing the operator from accidentally pressing an adjacent button. These devices are generally available for either horizontal or vertical button arrays. The dimensions shown are representative.

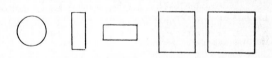

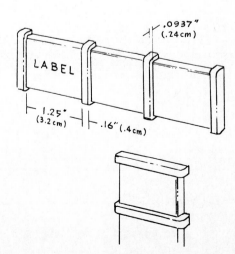

When inadvertent button operation could be serious, other, more accentuated guard systems should be considered. The one illustrated is effective because the height of the guard prevents the operator's finger from "slopping over" into the adjacent button area, and it also allows the labels to be seen with minimum interference.

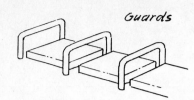

Push-button forces should be within the range of 10 to 20 oz (283 to 567 g) (although they can be as high as 40 oz, 1134 g) in order to reduce the possibility of inadvertent actuation.

Push-button action should be positive; i.e., there should be elastic resistance aided by slight sliding friction, starting slowly, building rapidly, and with a final sudden drop to indicate activation. An audible snap action is helpful under high-noise–level conditions, although the snap action must not be too heavy if the button is to be operated frequently over extended periods of time.

Contact Switches

Although not a true push button, the contemporary "contact switch" is often used to replace the mechanical push button. Since this type of switch provides no three-dimensional cue for the operator, it is important to provide extremely clear graphic delineation of each switch boundary.

It must also be made clear, by means of appropriate labels, which are switch functions and which are purely visual readouts, since both the switches and the readouts are flush with the panel surface.

Note: Contact switches should not be used where continuous, long-duration operations are expected, since it has been noted that operators tend to strike the push-button surfaces harder because of a lack of tactile or auditory feedback. Some complain of sore fingers after a long operating session.

TYPEWRITER KEYBOARDS

Typewriter Keyboard Operating Characteristics

Typewriter keyboards should be designed so that they can accept key strokes at a rate of about 20 per second, with short burst rates of up to 50 key strokes per second.

The force required to actuate the typical typewriter key should not exceed about 5 oz (142 g). However, to reduce inadvertent actuation, there should be at least 2 oz (57 g) of resistance provided. An adjustable force control is highly desirable, especially for mechanical machines.

Provide an acoustic shroud around the portion of the typewriter housing the principal noise-producing elements of the machine.

Provide a visible and intelligible centering display to aid the typist in identifying where the key will strike the sheet of paper, i.e., permanent indexing marks that will not be covered by the typewriter ribbon or other elements of the machine.

The ON/OFF control should be placed where it can be readily found, even by the inexperienced operator. Provide some type of indication to let the operator know when the machine is on and design the system so that, in the event the operator forgets to turn off the machine, it will automatically shut itself off after a given amount of nonoperating time.

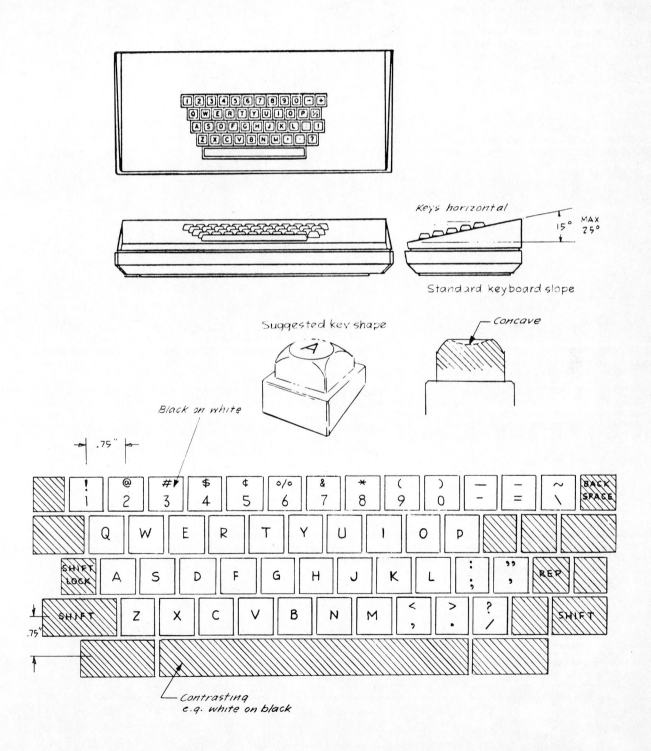

Keys horizontal

15° MAX 25°

Standard keyboard slope

Suggested key shape

Concave

Black on white

.75"

Contrasting e.g. white on black

CALCULATOR KEYBOARDS

Only the numerical keys on calculator keyboards have been somewhat standardized. Additional standardization would help the operator, at least for certain common functions.

The accompanying sketch illustrates suggestions for standardizing the common keying elements. The numerical keys are arranged in the order found on most calculators. The "clear" key is placed in the upper left-hand corner because this is the natural point to begin an operation; i.e., the operator first needs to make sure that the machine is clear for the first entry. The primary functions are arranged from bottom to top on the left of the numerical keys. Note that these are of a contrasting color. The "=" key is placed at the top of this series because the operator normally proceeds from this final operation to the display. Other functions should be separated from these more frequently used keys, and they should be in a contrasting color. Note that the "clear" and "=" keys are color-coded. Although the suggested colors are somewhat arbitrary, they do have significance; i.e., green suggests that the machine is ready, and yellow suggests caution before striking this final entry.

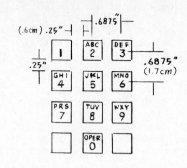

TOGGLE SWITCHES

Toggle switches come in a variety of shapes and sizes, most of which are satisfactory. One should, however, choose a toggle switch in which the "throw" or displacement is at least 30° so that the operator can tell at a glance which position the switch is in.

A toggle-switch handle should be at least 0.5 in (1.3 cm) long, and preferably no longer than 1.0 in (2.5 cm).

For smaller, shorter switch handles, the resistance of the switch should not exceed about 8 oz (227 g); for larger switches, it should not be more than 15 oz (425 g).*

The sketches at the right illustrate typical handle shapes that are generally available. The first is the most typical. The second is often representative of very small toggle switches. Such a shape is not desirable when the handle is long, however, because the user can be injured if he or she bumps against the handle. The third handle shown provides added tactile identification when a switch is used in the dark. The fourth handle is especially good, not only because it makes it easier to identify the switch position, but also be-

cause the special handle can usually be obtained in various colors, making it possible to color-code certain switches.

Switch separation is especially important when several switches are arranged side by side or vertically in columns. Note the different spacing requirements for vertical and horizontal arrangements.

Although three-position toggle-switch hardware is available, it is not recommended. It is not possible to maintain the 30° "throw" when more than two positions are provided. It is suggested that, when three or more switch positions are required, some other type of switch be selected.

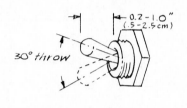

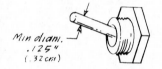

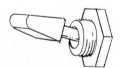

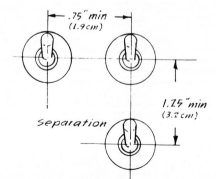

*Toggle switches should snap into place with an audible click.

Workstation Conceptualization and Design

Typically, a workstation includes several elements that should be spatially integrated in a manner that will support a total task. This requires careful consideration of the arrangement of elements so the task activities are simplified, grouped, arranged in sequence when appropriate, and minimally fatiguing due either to too much body movement or, equally bad in some cases, rigid confinement.

Dimensional features of the workstation should reflect an appreciation for human anatomic dimensions so that displays are placed where eyes can see them, controls located where hands *are* so that awkward reaches are eliminated, writing surface heights are optimized for ease of seeing and manipulating materials comfortably, and, finally, there is adequate clearance provided for knees and elbows.

The proper selection of support furnishings often is as important as packaging the individual electronic elements of a workstation. For example, a properly designed chair is often as important as the VDT package since the chair determines to a great extent how well a given operator "fits" the VDT setup. The amount and placement of accessory working surfaces are equally important for certain operational task situations. When support equipment must be accessible to the main workstation, furniture support for this must be compatible with the arrangement of the principal workstation components, e.g., communications equipment, printers, reference storage, etc.

NONOPERATOR FACTORS

Several nonoperator considerations are important in the design of equipment packaging.

Often, in our enthusiasm to create the ideal operator interface, other user interface problems are forgotten. Perhaps most important is the ease of access for testing, servicing, and repairing an equipment item. In conjunction with this aspect of maintenance, mobility of equipment should be kept in mind. Incidentally, ease of access should include not only the physical aspect, but also the organizational features and labeling relating to troubleshooting. In this case, a maintenance technician is in a sense an "operator." Most of the good human engineering design principles established for operator interfaces apply equally to maintenance interfaces.

It should also be noted that equipment package mobility is often important in setting up the basic work station, especially when equipment elements may be moved frequently from one operation to another.

WORKING POSITION

Workers or operators should stand when:

1. The task they must perform requires reaching beyond a point that typically can be reached comfortably while seated.
2. They have to move frequently from one type of task or workplace to another.
3. Certain tasks require lateral reaching beyond a specific task site, i.e., when it is easier to step to another position.
4. They frequently interface with others who are standing.
5. Several persons have to work jointly at a large display area such as a diagram.
6. They have to reach long distances to touch, transport, or adjust controls.

Workers or operators should be provided an alternative sit-stand configuration when:

1. They normally remain at a task more than 30 min, but others frequently must observe their operation.
2. They need the advantages that seating provides for their major task, but often have to move to another position for a temporary standing task.

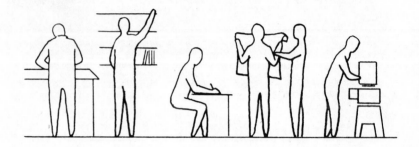

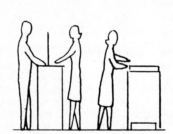 

Have workers stand if tasks require frequent position change and tasks are short.

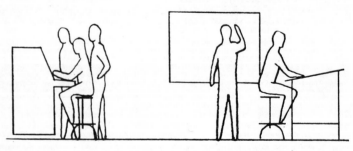

Provide sit-stand configuration if primary task is long, but frequently interspersed with standing operations — getting up and down is fatigueing.

Workers or operators should sit when:

1. They must work at the same task for an extended time period (30 min or more).
2. Task precision demands steadiness both for them and for the devices being used.
3. They need to be restrained to prevent their displacement by dynamic environmental forces.
4. They have to use their feet to operate controls.
5. They are required to apply maximum force to operate a control.
6. They have to perform extensive writing tasks.

Note: Although the above principles apply generally, it is obvious that each decision must be based on a thorough analysis of all the interacting factors, e.g., restraint, steadiness, mobility, interoperator coordination, and potential fatigue. For example, a draftsman usually can perform fairly precise drawing movements either standing or sitting. On the other hand, a vehicle operator generally cannot perform the typical hand and foot control operations as well when standing as when seated.

In addition, the nature of the equipment configuration may require that elements be widely spaced, in which case excessive reaching can be minimized by providing a more mobile operator facility.

Above all, avoid the tendency to repeat traditional workplace configurations without making sure these are compatible with all the task demands.

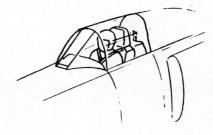

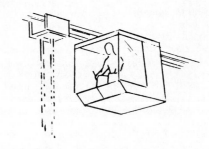

GENERAL WORKER AND OPERATOR POSITIONING CONSIDERATIONS

1. Avoiding Awkward Positioning
a. Minimize the necessity for operators to lean; i.e., arrange the workplace so that operators can maintain a more or less normal alignment of their bodies, especially when standing.
b. Operators should not have to use their maximum reach capability; i.e., they should be able to shift their bodies to a better position.
c. Operators should not have to sit or stand for long periods with their head, torso, or limb positions skewed; i.e., they should be able to keep their heads, necks, torsos, and limbs in a symmetrical relationship as much as possible.
d. Operators should not be forced to work frequently or for long periods with their hands and arms above normal elbow level.
e. Avoid positioning operators in supine or prone positions.
f. If the limbs must be extended for long periods, provide some kind of limb support.
g. Configure the workplace so that operators can see what they are doing without having to assume an awkward head or body position to see around their hands or the control device.

2. Creating Arrangements That Are Compatible with Normal Limb Movement Patterns and Reach Limitations
a. Minimize the need for operators to move their torsos during continuous control operation, except where a ballistic advantage is desirable.
b. Repetitive arm and leg motions should be in the direction that limbs articulate normally about typical shoulder, elbow, wrist, hip, knee, or ankle pivot points.
c. When both hands or both feet are involved simultaneously, create a configuration in which the motions are opposite rather than in the same direction; i.e., the right hand should rotate in the opposite direction to the left, and the right foot should push forward while the left relaxes in the aft direction, as in aircraft rudder pedal movement.
d. Minimize the necessity for operators to shift their position in a seat in order to look at a display or operate a control. In certain situations, however, it may be desirable to provide seat movement, i.e., fore and aft, laterally, or rotating.
e. Avoid workplace arrangements that force operators to stand too close to, or lean across, a hazardous element, such as a moving belt or chain or a hot component.
f. Take into account the extra clearance requirements and/or reduced mobility caused by special apparel that operators may be required to wear and make sure that there are no features within the mobility envelope that could prevent proper limb movement, i.e., components which might be bumped, which limit full limb movement, or upon which clothing might become snagged.

3. Force Application
a. Place manipulable objects or controls in positions that are compatible with the best geometric relationship for applying pushing, pulling, or rotating force by operators whenever such forces cannot be minimized.
b. Provide an appropriately positioned support against which operators can minimize counterforce effects, e.g., an armrest, handrest, or backrest.
c. Arrange force-demanding controls in positions where operators can apply the necessary force without disturbing their normal body position, especially when this might interfere with a primary visual or control activity and/or with necessary body-referencing posture.

4. Minimizing Fatigue-Producing Workplace Arrangements
a. Provide backrests for seated operators.
b. Provide armrests.
c. Provide handrests when operators are using a continuously operated controller, such as a desk-mounted joy stick or roller-ball controller.
d. Provide footrests.*

5. Minimizing Potential Safety Hazards
a. The operator's posture should be such that his or her body is in a position that can accommodate dynamic force inputs, such as excessive spinal loads during deceleration, the impact of a sudden landing, or the buffeting of a vehicle as it impacts against rough terrain or ocean waves.
b. Provide body restraints for vehicle operators which will protect them from impacting the interior structure or equipment and which will not allow them to slip into a position in which the restraint itself, such as a seat belt, becomes a hazard.
c. Make the workplace as safe as possible in terms of locating and designing components so that they will not penetrate an operator's skin upon contact, i.e., so that they will spread the energy rather than concentrate it.

KEY FACTORS TO CONSIDER FOR EACH TYPE OF OPERATOR WORKING POSITION

1. Standing Operators
A smooth, level surface should be provided.
There should be sufficient surface area for operators to establish an adequate spread of their feet, to move when necessary, and to brace their feet when required.
A nonslip surface should be provided if a platform is subject to movement, as in a vehicular environment.
A resilient surface should be provided if operators must stand all day.
Visually observed workplace elements should be arranged so that they can be seen without excessive movement, so that there is no

*Refer also to the discussion of problems of awkward positioning.

parallax for reading instruments, and so that reading distance and size of display are compatible with readout precision.
Manipulative tasks should be arranged and/or oriented to be compatible with respect to reach convenience, required motion patterns and excursions, force application requirements, precision demands, and response speed. Special attention should be given to the implications of eye-hand coordination.

2. Sit-Stand Operators
The eye reference for both seated and standing operators should be approximately the same.
The arrangement of visual displays that have to be monitored by both seated and standing operators should be such that there is the same level of readout accuracy from both positions; i.e., operators should not make reading errors because of parallax.
If both seated and standing operators may have to use the same control device, place this in a position that will minimize interoperator interference.
Provide a footrest for seated operators.

3. Seated Operators
Provide a seat that ensures optimum working posture for the tasks being performed.
Arrange visual displays and controls so that an operator's hand will not cover critical displays and so that the displays are normal to the expected viewing line of sight.
If the primary viewing task is directed outside the workplace (vehicle control), position the most important internal displays so that they can be viewed without excessive eye movement from the nominal exterior line of sight.
Position all controls so that, in manipulating them, operators do not appreciably move their nominal eye reference and possibly miss seeing important events occurring outside or on the principal internal display.

4. Miscellaneous Considerations
a. Organizational standardization versus individual adjustment: When similar workplaces are to be repeated, a compromise should be sought between the benefits of a standardized workplace (which permits operators to shift from one workplace to another without confusion) and the benefits that accrue from allowing individual operators to modify their workplace to fit each type of activity more conveniently (i.e., to rearrange tools, materials, or other aids so that they are more accessible).
b. Illumination: Do not assume that a generalized type of illumination will satisfy all task-seeing requirements; i.e., the type and location of light fixtures should be determined according to the varying needs of the operator. Especially important is the creation of seeing conditions that minimize glare and reflection problems.
c. Storage: Almost all workplaces require the provision of storage space for books, files, tools, or writing and drawing equipment and materials. Equally important is the provision of storage space for communications devices, discarded items,

and personal articles such as glasses or eye shields.

d. Service and maintenance: Careful consideration should be given to ensuring that, in arranging the workplace and/or packaging of equipment, access to critical elements for service or other maintenance is provided.

e. Design for handicapped operators: Whenever a workplace may be used by a person in a wheelchair, a blind person, or a deaf person, special care must be taken not only to dimension the workplace properly but also to provide those special features necessary for the handicapped operator to perform more effectively within his or her limitations (e.g., tactile features for the blind and special visual signals for the deaf).

f. Environmental considerations: Be aware of the possibility that a particular individual workplace may reduce the effectiveness of general ambient thermal, ventilation, and noise control systems, thus requiring additional features to ensure that the operator does not become overheated or is not subjected to respiratory contamination or to interfering or annoying noises that may be produced within his or her particular workplace.

g. Manipulative clearance for large materials: Often large materials must be maneuvered into place before they can be worked on. Provide adequate clearance so that such materials can be maneuvered into place easily.

BASIC TASKS IN RELATION TO WORKPLACE DIMENSIONING

1. Writing: The writing surface should be at approximately elbow level and should be relatively horizontal (maximum 5° slope); greater slopes make it difficult to keep a pen or pencil from rolling off the surface.
2. Typing: The center of the keyboard should be at about elbow level.
3. Data entry keyboard: The center of the keyboard should be at about elbow level. An associated visual display should be positioned so that the operator's line of sight is perpendicular to the face of the CRT.
4. Drawing: A drawing board used for precision work should slope approximately 3° to 4°; this provides the best compromise between reaching, viewing, and precise instrument manipulation. For less precise drawing, sketching, or artwork, the drawing surface should be adjustable, so that the artist can match the size of the work with the best position for sketching, painting, etc. A sit-stand arrangement is most desirable for average mechanical drafting work since it allows the worker to reach all areas of a large drawing easily.

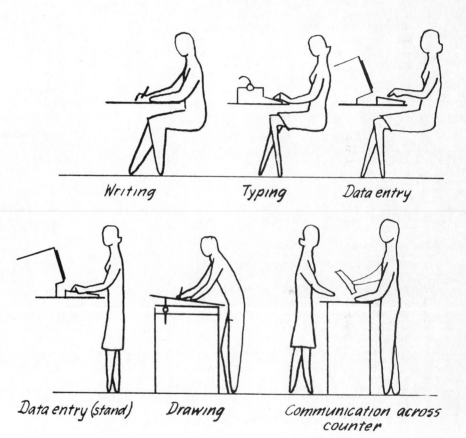

Writing   *Typing*   *Data entry*

Data entry (stand)   *Drawing*   *Communication across counter*

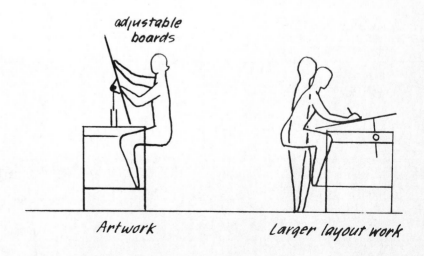

adjustable boards

Artwork   *Larger layout work*

5. Storing and retrieving shelves: Design and/or select storage shelving that allows the worker to reach and grasp objects to be stored or retrieved easily and without fear of dropping them. Avoid deep, high shelving since it is easy for small objects to be pushed to the rear of a shelf, where they are neither visible nor manually accessible.

6. Office machine use: Either select machines that have been properly designed (including special stands) or select or build tables that assure proper positioning of the principal work level of the machine. The work level should be at approximately the operator's elbow height.

7. Tool operation: Various tools require that the work level be carefully chosen, especially if there is a requirement for precision and/or the application of controlled force.

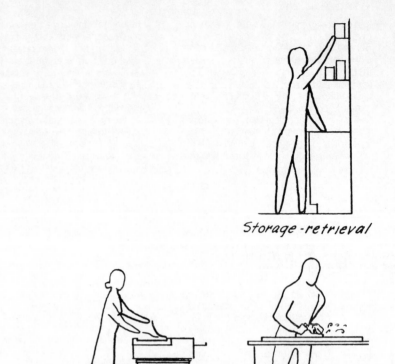

Storage-retrieval

Office machine operation

Shop work

Nominal Guideline Dimensions

Although the accompanying general dimensional suggestions are recommended to approximate the optimum, one should be careful not to assume that they necessarily apply exactly to a specific problem. Use of a test mockup to verify these is recommended.

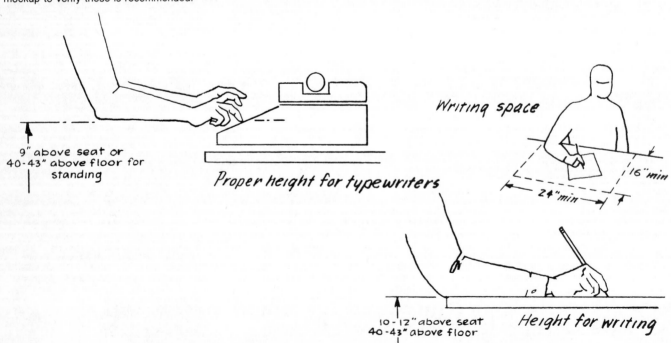

9" above seat or 40-43" above floor for standing

Proper height for typewriters

Writing space

24" min

16" min

10-12" above seat 40-43" above floor

Height for writing

ELECTRONIC CONSOLES AND RACK-MOUNTED PANELS

1. Seated Operator Consoles

Operators who are seated at an electronic console are necessarily limited in terms of how far they can reach. In addition, their viewing angles may be critical in order that display faces are reasonably perpendicular to their normal lines of sight.

In general, consider the suggestions illustrated by the accompanying sketch:

a. Position the primary display—say, a large CRT—slightly below the operator's eye level, but sloped to match his or her normal head slump of about 10°.

b. Adjustment controls that are used most frequently should be positioned at approximately elbow level.

c. If a finger joy stick or roller-ball controller is required, place it nearer the back edge of the desk, not near the outer edge, where the operator will not have a resting surface for his or her arm.

d. Displays or controls that are used infrequently can be located in less convenient positions, as long as they are not "detached" from related functional monitoring areas.

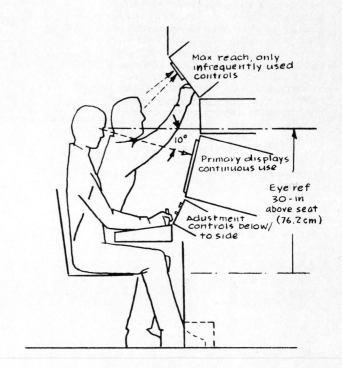

Max reach, only infrequently used controls

10°

Primary displays continuous use

Eye ref 30-in above seat (76.2 cm)

Adjustment controls below/ to side

2. Standing Operator Cabinets

See the accompanying sketch.

3. See-Over Consoles

If the operator needs to see over the console to monitor other displays, reduce the console height so that the operator is not required to stretch his or her torso or neck.

secondary displays & controls

primary displays, controls, tape reels

60" (152.4 cm)

seldom use controls no displays

No operating displays or controls

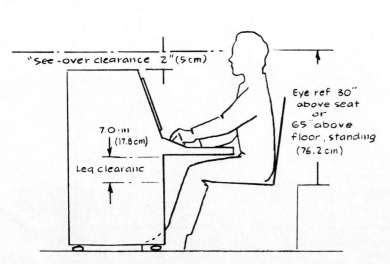

"See-over clearance 2"(5cm)

7.0-in (17.8 cm)

Leg clearance

Eye ref 30" above seat or 65" above floor, standing (76.2 cm)

4. Extreme See-Over Configurations

When it is necessary for an operator to "see down" (as illustrated in the accompanying sketch), one should examine the total facility architecture in conjunction with the design of the operator's control station. As the illustrative example implies, the operator's position for viewing objects outside the workplace has to reflect consideration for *(a)* what the operator has to see, i.e., the required down angle; *(b)* the operator's position relative to the structural obstruction; and *(c)* the operator's position in relation to the shape and size of the control package so that the package configuration (at least) allows the operator to take full advantage of whatever clear view the structure allows. In some cases it may be desirable to use a sit-stand configuration in order to take advantage of the fact that the operator is freer to lean over the console.

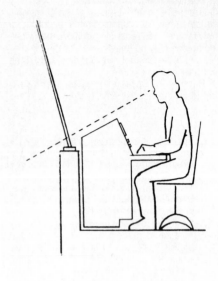

5. Large Plot Subsystems

The two accompanying sketches illustrate general dimensional guidelines for large plot subsystems; these are based on *(a)* limitations created by knee clearance if the operator sits and *(b)* convenient use area when the operator stands.

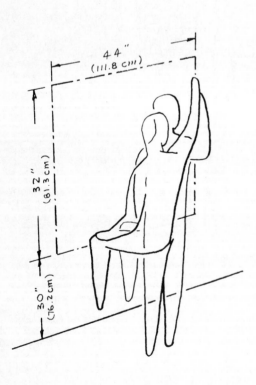

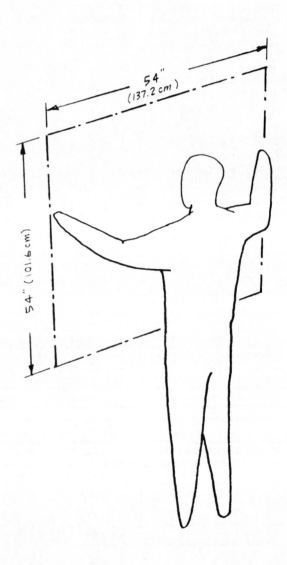

Other Considerations

As a general rule, panels for either seated or standing operation are easier to use when the displays are laid out for horizontal, sequential scanning. This is due to the fact that an operator's eye motions and head movements are less tiring in the horizontal direction. Studies also have shown that operators can respond slightly faster when they attempt to read instruments or other visual information from horizontally arrayed data than when this is presented in vertical arrays.

On the other hand, standing operators, because they are free to walk backward and to the side, can more easily adjust to control and display arrangements that are organized in vertical patterns.

Horizontal arrangement for seated operation.

Horizontal and vertical for standing operator.

Display Positioning Relative to Illumination and Reflection or Glare

Large CRTs and/or glass-covered instruments are subject to interference by ambient lighting conditions. Whenever possible, one should attempt to control both the positioning of displays and the ambient lighting conditions. Although an important objective usually is to position a display so that the face is perpendicular to the operator's normal line of sight, it may be necessary to compromise in order to eliminate undesirable reflections from artificial and/or natural light sources.

When control of ambient lighting is not the responsibility of the designer, other steps must be taken to minimize the effects of possible glare or reflection. The glare shield illustrated in the accompanying sketch is typically used to accomplish this.

Arranging for Added Convenience

Although the typical console or equipment operator has considerable reach capability, it is desirable to minimize the necessity for reaching, especially if it may be required on a frequent basis. The so-called wraparound concept is helpful as long as care is taken to ensure that unusual interferences are not introduced.

The wraparound console is especially helpful for the seated operator, as shown in the accompanying illustration. If the seat has to be secured to the floor (as it does aboard ship), a careful analysis of reach limits must be made. If a movable seat is acceptable, the arrangement can be expanded, taking advantage of a chair with casters.

A sit-stand operation (see the accompanying sketch) that includes both worktable and panel rack operations can be accommodated best if the rack can be made movable.

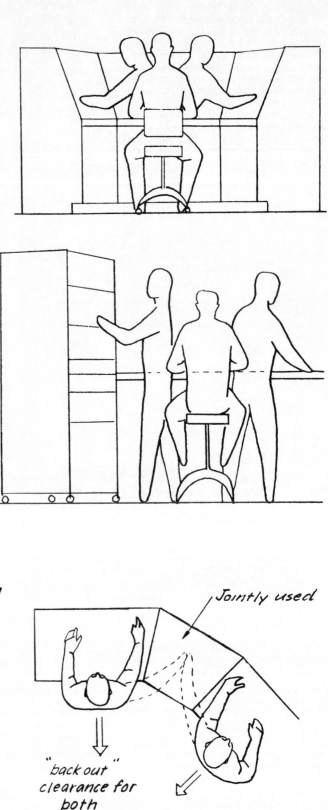

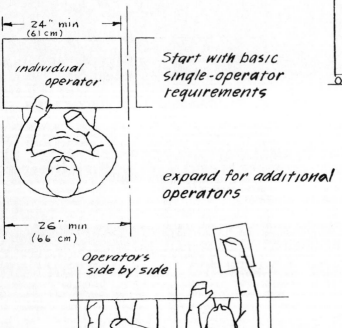

24" min (61 cm)

individual operator

26" min (66 cm)

Start with basic single-operator requirements

expand for additional operators

Operators side by side

30" (76.2 cm)   30"

Jointly used

"back out" clearance for both

**GUIDELINES FOR POSITIONING
CRT-TYPE DISPLAYS**

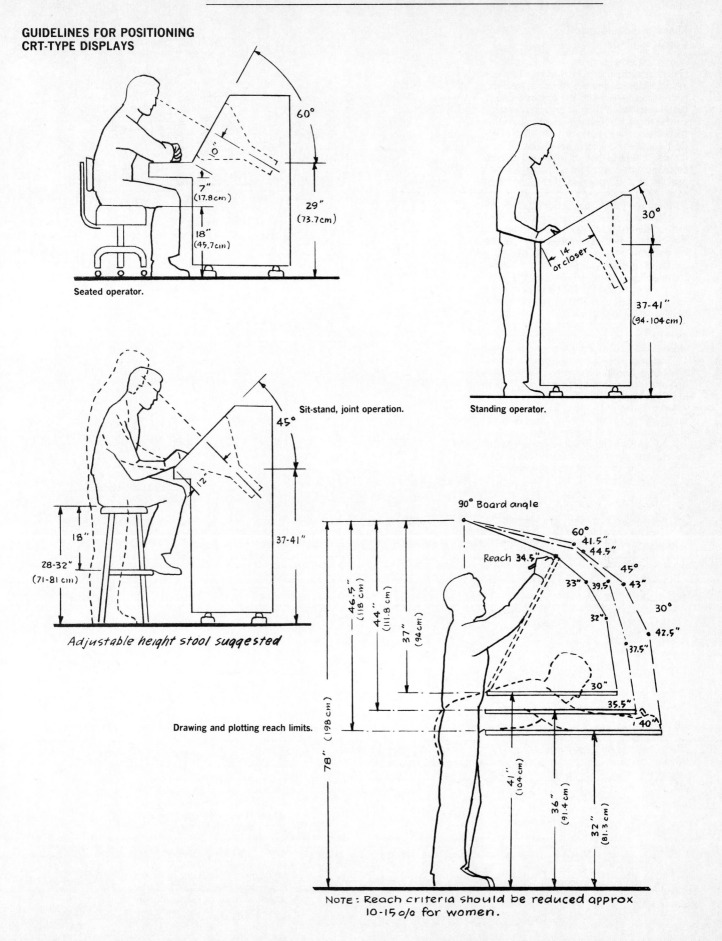

Seated operator.

60°

7"
(17.8 cm)

29"
(73.7 cm)

18"
(45.7 cm)

Standing operator.

30°

14"
or closer

37-41"
(94-104 cm)

Sit-stand, joint operation.

45°

12"

37-41"

18"

28-32"
(71-81 cm)

Adjustable height stool suggested

Drawing and plotting reach limits.

90° Board angle

60°
41.5"
44.5"

Reach 34.5"

45°

33"   39.5"   43"

30°

32"

42.5"

37.5"

30"

35.5"

40"

46.5"
(118 cm)

44"
(111.8 cm)

37"
(94 cm)

78" (198 cm)

41"
(104 cm)

36"
(91.4 cm)

32"
(81.3 cm)

NOTE: Reach criteria should be reduced approx
10-15 o/o for women.

Miscellaneous Console Considerations

Avoid the tendency to make all elements of a console operator's operation fixed. That is, it is sometimes desirable to modularize portions of the console so that the modules can be rearranged to be more compatible with the particular activity. One typical example is the separate packaging of a keyboard. The operator may perform several different operations at the console, only one of which requires the keyboard. If the keyboard is fixed in a central position, it is obvious that that prime space cannot be used in any other manner.

When a pencil joy stick or track-ball controller is involved in the console use, be sure that the control position is far enough forward so that the operator has a place on the desk for his or her arm.

CRT viewing angle is important in that when the device is used for long periods, the operator should not have to hold his or her head in an awkward position. Although we often try to set the CRT angle in a so-called compromise position, this may still create problems for either the shortest or the tallest operator. One obviously desirable approach is to make the CRT package adjustable, as shown in the accompanying sketch. Not only does this allow operators to find the best vision angle to reduce neck fatigue, but it also allows them to adjust the face of the display to minimize any annoying reflections of themselves or of objects behind or above them in the room.

A convenient and effective workplace can be created using independent modules, as shown in the accompanying illustration.

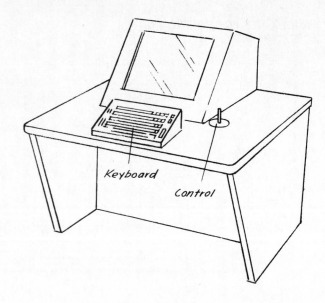

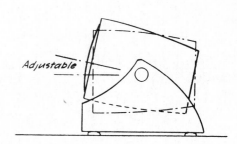

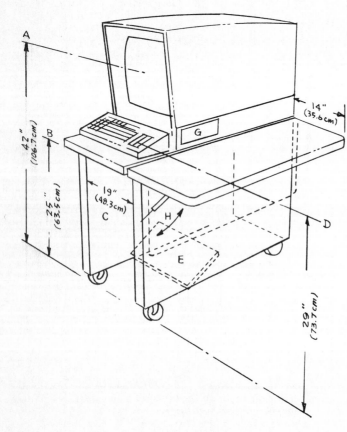

A—The center of a vertically oriented CRT should be approximately as shown. Note that the case extends over the tube to minimize glare from overhead ambient light.

B—Vertical knee clearance minimum.

C—Lateral knee clearance minimum.

D—Center height of keyboard.

E—Footrest.

F—Minimum width for fold-down writing surface.

G—CRT controls (under cover).

H—Supporting bracket for fold-down desk. Casters with locks allow the unit to be moved around to fit the user's needs.

VEHICLE OPERATOR WORK STATIONS

As the accompanying illustrations show, the vehicle operator's position relative to external and internal viewing requirements depends on the size and shape of the proposed vehicle:

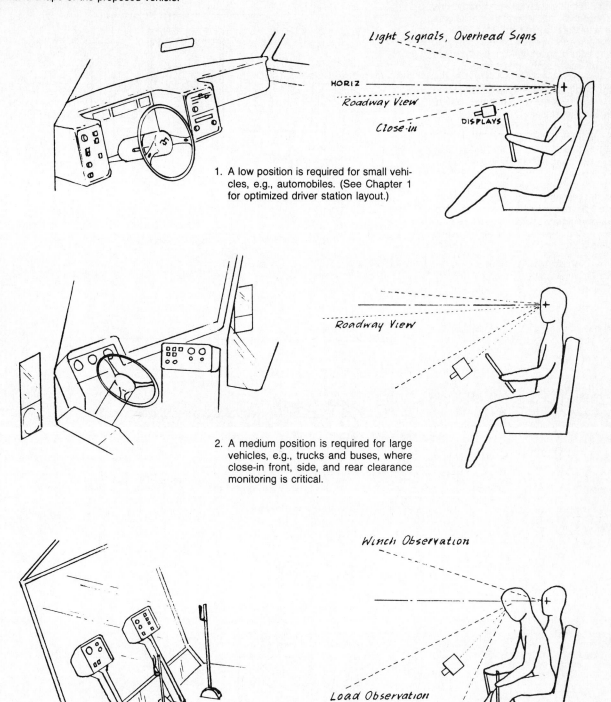

1. A low position is required for small vehicles, e.g., automobiles. (See Chapter 1 for optimized driver station layout.)

2. A medium position is required for large vehicles, e.g., trucks and buses, where close-in front, side, and rear clearance monitoring is critical.

3. A flexible position is required for special vehicles, e.g., construction machines, in which observation of external components and related materials, pedestrians, or terrain is vital.

LARGE GROUP DISPLAY AND OBSERVER ARRANGEMENT

When a number of operators must view a large visual display in conjunction with console operation, the large group display should be arranged so that the various observers can see the display from their normal working positions without having to peer around their neighbors or their neighbors' equipment. Many alternative arrangements are, of course, possible, and sometimes a compromise must be made because of facility space constraints.

The accompanying illustrations and guidelines are helpful in making sure that all the sight-line requirements have been evaluated.

As the two accompanying sketches illustrate, angle of view may be improved by seating the operators further apart.

Operators and their consoles can be staggered in order to maintain the recommended screen-observer line-of-sight limit of 60°, as shown in the accompanying illustration.

When the recommended clear line of sight cannot be accomplished by lateral spacing, consider placing the operators in the rear on platforms so that they can see over the heads of those in front.

Possible but awkward

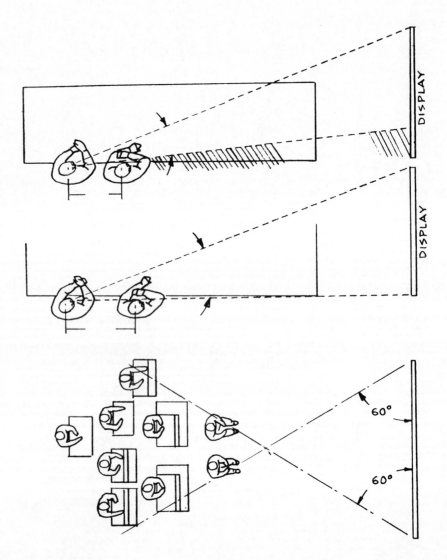

If, because of space constraints, the large display cannot be placed far enough away from a group of observers, consider breaking up the display so that the edges can be rotated for a more suitable angle of viewing.

When the large display is created by a projector, take care to arrange the projector and the observers so that intervening observers do not block the projected image.

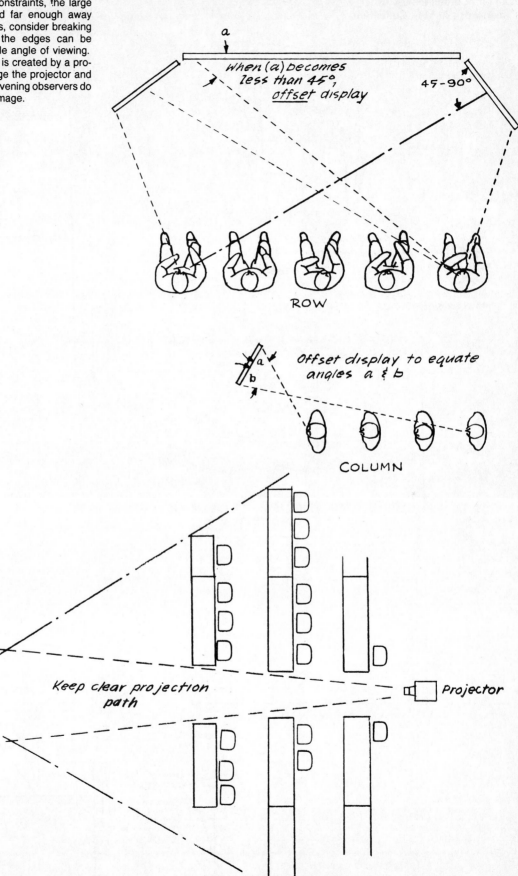

**Miscellaneous Guidelines for Other
Typical Operator Working Conditions**

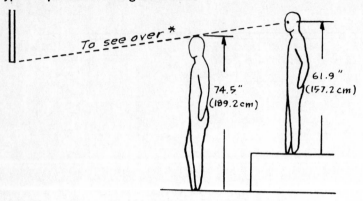

*NOTE: *Values for male observers. For mixed male/female situations, consider tall male in front, short female in rear.*

Manual accessibility and clearance.

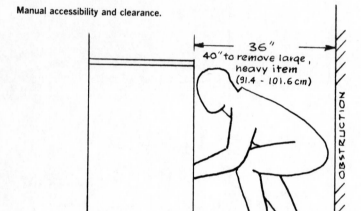

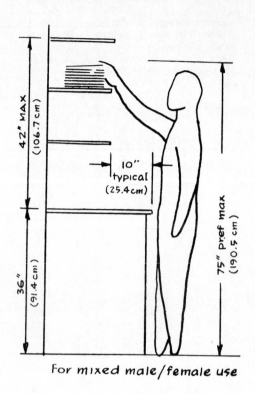

For mixed male/female use

Visual accessibility.

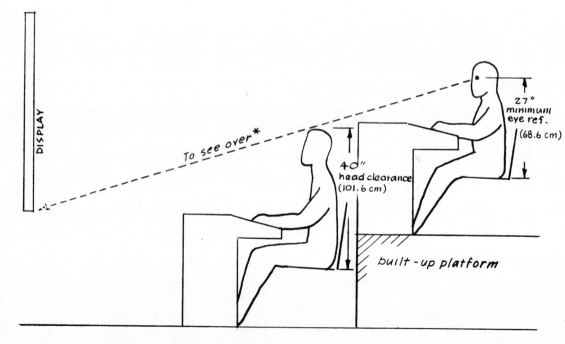

Television Viewing

Screen Size*	Minimum Viewing Distance	Maximum Viewing Distance
17 in (43 cm)	4 ft 11 in (1.47 m)	14 ft 9 in (4.43 m)
19 in (48 cm)	5 ft 1 in (1.53 m)	15 ft 2 in (4.58 m)
21 in (53 cm)	6 ft 4 in (1.89 m)	19 ft (5.7 m)
23 in (58 cm)	6 ft 6 in (1.95 m)	19 ft. 4 in (5.8 m)
25 in (64 cm)	7 ft 6 in (2.25 m)	21 ft. 9 in (6.5 m)

*Diagonal.

For arranging observers in direct TV viewing situations, use the guidelines shown in the accompanying table. For projected television viewing, use the following guidelines:*

W = Screen width

Ambient light level should be between 0.1 and 2.0 ftc for comfortable viewing.

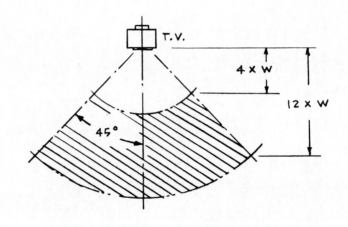

T.V.

4 X W

12 X W

45°

Preferred viewing area for watching T.V.

1. Screen brightness   2 to 20 ft (0.6 to 6.1 m) L
2. Brightness ratio   2:1—excellent
     3:1—very good
     10:1—acceptable
3. Contrast ratio   100:1 for pictorial images
     25.1 for printed characters
     5.1 for white characters on black background

*A minimum contrast ratio of 30:1 is for the poorest conditions, e.g., the observer who is furthest away, high ambient room lighting levels, and poor materials.

**Typical Military Operator Dimensional
Guidelines***

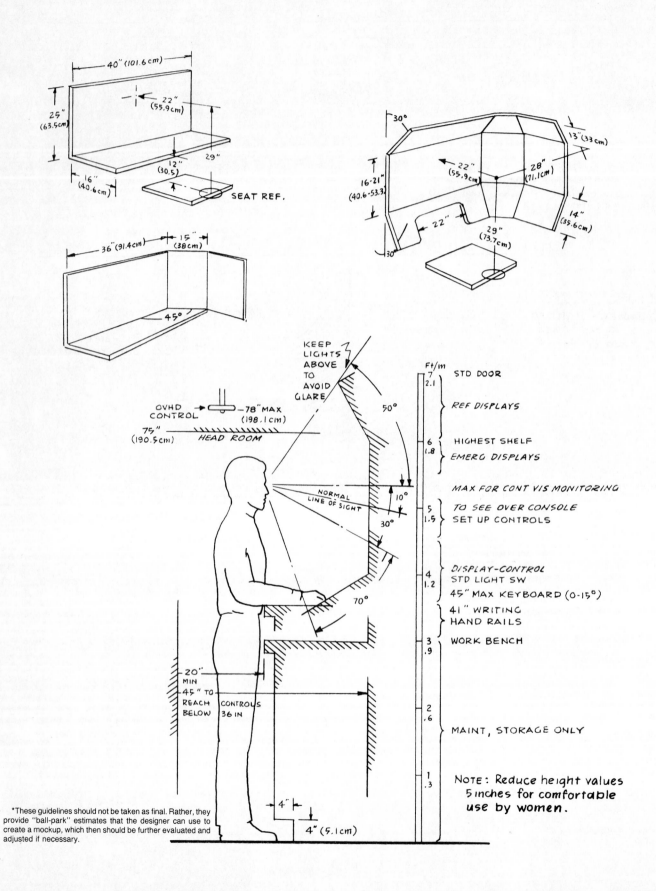

*These guidelines should not be taken as final. Rather, they provide "ball-park" estimates that the designer can use to create a mockup, which then should be further evaluated and adjusted if necessary.

MAINTENANCE WORKPLACES

Although equipment maintenance often is not considered as important as the operator's interface, design of the maintenance interface is important to the maintenance technician and therefore should be considered at the same time that operator interfaces are being designed. Several typical examples of maintenance problems, created by lack of proper attention, are illustrated in the accompanying sketches.

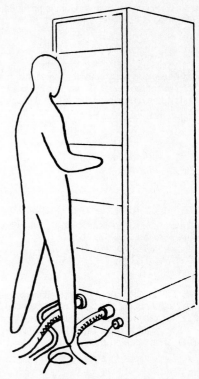

Electrical cables in the way.

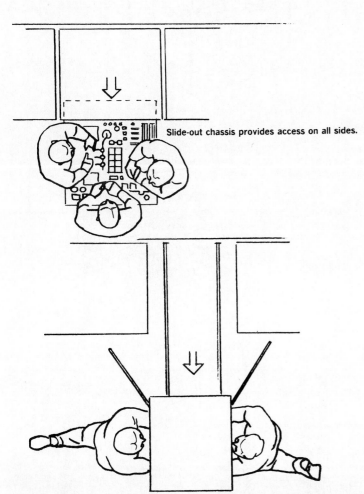

Slide-out chassis provides access on all sides.

Cabinet tracks allow the cabinet to be removed from among a row of cabinets that block access for maintenance.

Should exit the back or top of equipment.

Different equipment packaging concepts either aid or hinder the maintenance technician, as illustrated by the accompanying sketches:

Access is better if the case can be lifted off the chassis. More freedom to work is provided, and there is no cover to prop up.

Avoid designs in which the chassis has to be lifted out of the case or in which a cover has to be removed or propped out of the way.

Design covers so that the minimum number of fasteners have to be removed.

Fasteners should be captive so that they do not become mislaid or lost.

When high voltages are involved, removal of the case or cover should automatically shut the power off. A secondary switch or jumper should be used to reactivate the power when maintenance is being performed.

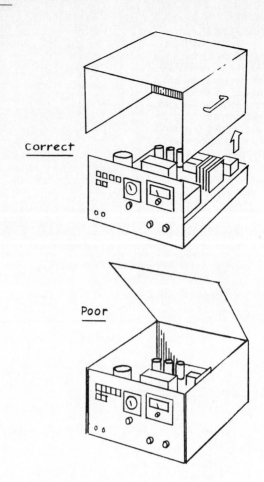

Correct

Poor

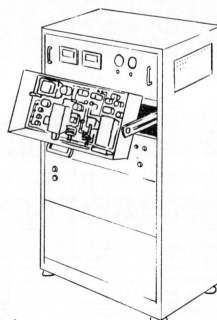

A rotating chassis makes both the top and the bottom accessible.

The fold-out package makes the entire inner assembly accessible.

Closed

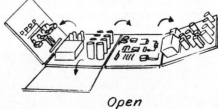

Open

EQUIPMENT AND CABINET DOORS

Cabinet doors and doors for gaining access to equipment and home appliances should be designed and located to simplify the user's problem of getting the doors open and holding them open while he or she is working or putting something inside, and they should be planned to "fit" the constraints of the expected range of installations. The accompanying sketches illustrate a few of the common problems associated with typical door designs.

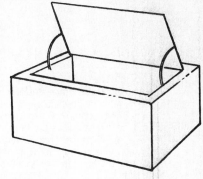

A "lid-type" door should be hinged to prevent the cover from being lost, and guides should be provided so that there is a means for holding the cover open. The same objectives pertain to the design of doors for a typical office storage cabinet.

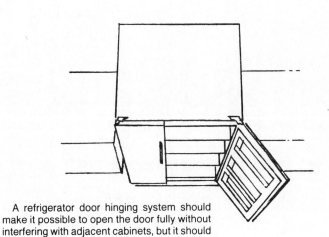

A refrigerator door hinging system should make it possible to open the door fully without interfering with adjacent cabinets, but it should be designed to close automatically (slowly, of course).

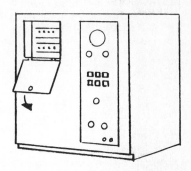

A maintenance or adjustment compartment door should be hinged as shown so that the cover will remain open while adjustments are being made.

Typical compartment closure schemes for entertainment cabinets are acceptable as long as the doors do not bind. The handles should be designed so that the user's fingers are not pinched.

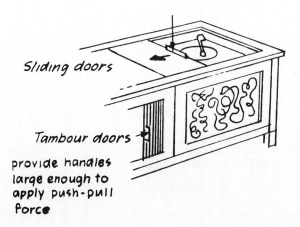

Sliding doors

Tambour doors provide handles large enough to apply push-pull force

Injury-Causing Door Features

The following common errors in design may
cause considerable difficulty for the user.

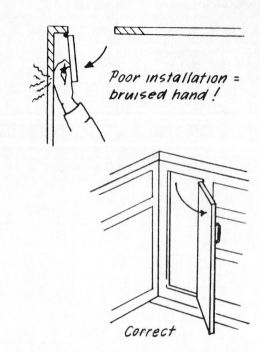

*Poor installation =
bruised hand!*

Correct

1. Incorrect Door Hinge Pattern

Equipment or cabinet doors should not be
mounted so that the user's hand might be injured if he or she opens the door too far. In
some instances the user may be applying considerable force in order to overcome the re-
striction of a door clasp; when the clasp suddenly lets go, the user's knuckles may be
cracked on an adjacent structure before he or
she can reduce the amount of force being applied.

Never locate the handles of adjacent doors
so that they could coincide during an opening
procedure.

2. Sharp Corners

Although the cost of eliminating sharp corners
may not seem warranted to the manufacturer,
the typical equipment user is not very sympa-
thetic to the so-called cost-effective sharp corner that may just have punctured his or her
hand. Be especially alert for sharp corners
that people might inadvertently hit with their
heads.

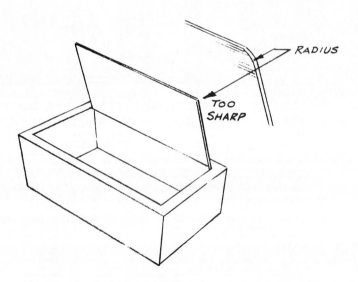

RADIUS

*TOO
SHARP*

*leave
clearance*

3. Sliding Doors

Provide stops so that people will not pinch
their fingers as they slide a door against another part of the cabinet. At least 1½ in (3.8
cm) of clearance is needed to provide finger
protection.

Equipment Door Size

Equipment door size obviously relates to the specific activities that may take place as technicians put their hands and/or an object through the opening. The door size should permit the following:

1. Using a common screwdriver with freedom to turn the hand through 180°

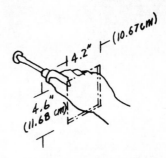

2. Using pliers and similar tools that require gripping

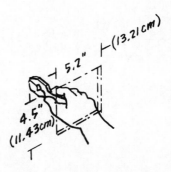

3. Using a T-handle wrench with freedom to turn the tool and hand through 180°

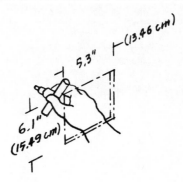

4. Using an open-end or box-end wrench with freedom to turn the wrench through at least 60°

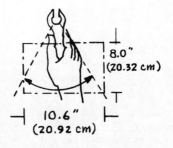

5. Grasping and manipulating small objects (up to 2¼ in, 5.7 cm wide) with one hand

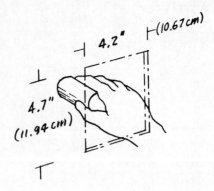

6. Grasping large objects with one hand

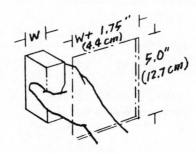

7. Grasping large objects with two hands, with the hands extended through the opening up to the length of the fingers

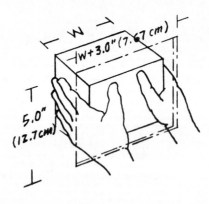

8. Grasping large objects with two hands, with the arms extended through the opening

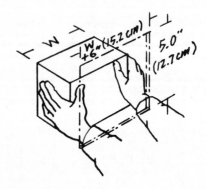

GUIDELINES FOR THE DESIGN AND SELECTION OF FASTENERS

Use the minimum number of fasteners compatible with requirements for securing components. In designing equipment packages and covers, use tongue and slot or similar techniques to help reduce the number of fasteners required.

Use "captive" fasteners to avoid loss of all or part of the fastening device.

It should be readily apparent when a fastener has been "released."

Work space should be provided around a fastener for fingers, tools, or "wrenching space."

Standardized fasteners should be used wherever possible to minimize stocking requirements and the number of different tools required.

Use manually operated fasteners where suitable to avoid obtaining or carrying tools.

Use fasteners that are operable by commonly used rather than special tools.

Use the same type and size of fastener consistently for a given application.

Screws, bolts, or nuts with different threads also should have clearly different physical sizes to minimize their being interchanged.

Consider how a worn or stripped fastener can be removed before selecting it; avoid stud fasteners that are an integral part of the machine or housing. Use screws rather than rivets.

Select safety fasteners for applications where vibration could result in loosening and perhaps loss of the fastener during operation.

Fasteners that are operable by hand should also be rugged enough to withstand operation by some type of tool.

Spring-load catches so that they lock on contact, rather than requiring a separate locking step.

Use "long latch" catches to minimize the possibility of inadvertent latch release.

When a latch has a handle, locate the latch release near the handle so that it can be operated with one hand.

Screw heads should have deep slots to reduce tool slippage and damage to the screw head.

Place the screw head so that offset screwdrivers are not required.

When bolts and nuts are used, make sure there is clearance to both ends.

Keep bolts as short as possible so that they will not snag personnel or equipment. Do not locate bolts where inadvertent replacement by a longer bolt might interfere with the movement of an adjacent mechanism or puncture adjacent cover or "skin."

Coarse threads are preferable to fine threads for low torques because they reduce the possibility of cross threading.

Avoid the use of left-hand threads unless the system requirements demand them; then identify both bolts and nuts clearly by a suitable marking, shape, or color.

Cotter keys or pins should fit snugly, but they should not have to be "driven out."

Use safety wire only when self-locking fasteners cannot withstand the expected vibration or stress. Attach safety wire so that it is easy to remove and replace.

Use retainer rings which hold with a positive snap action. Avoid rings which become difficult to remove when they are worn.

Provide retainer chains to capture fasteners which have to be completely removed but which could easily become lost.

Avoid the use of fastener systems in which a captive spring could suddenly drive the fastener into the face of the technician once the fastener is released or in which the spring itself pops out and becomes lost. Similarly, avoid systems in which a washer easily becomes detached and lost without the technician's knowing about it. Avoid springs that are so strong that a technician has to use a special tool to depress the spring before engaging the basic fastener.

FASTENER, TYPES AND DESIGN
CONSIDERATIONS*

Type	Description
	Adjustable pawl fastener. As knob is tightened, the pawl moves along its shaft to pull back against the frame. 90° rotation locks, unlocks fastener.
	"Dzus"-type fastener with screwdriver slot. Three-piece one-quarter–turn fastener. Spring protects against vibration. 90° rotation locks, unlocks fastener.
	Wing head. "Dzus" type. 90° rotation locks, unlocks fastener.
	Captive fastener with knurled, slotted head. Retaining washer holds the threaded screw captive.
	Draw-hook latch. Two-piece, spring latch, base unit and striker. When engagement loop is hooked over striker, depressing lever closes unit against force of springs. Lever is raised to unhook.
	Trigger-action latch. One-piece, bolt latch. Depressing trigger releases bolt, which swings 90° under spring action and opens latch. To close, move bolt back into position.
	Snapslide latch. One-piece snapslide. Latch is opened by pulling lever back with finger to engage release lever.
	Hook latch. Hook engages knob on striker plate. Handle is pulled up locking in place. To release, reverse procedure.

*Human Factors Engineering Design for Army Materiel, MIL-HDBK-769, 1975.

Types of Fasteners

Type	Description	Maintainability Considerations	Approximate Operating Time
* *Knurled*	Adjustable pawl fastener. As knob is tightened the pawl moves along its shaft to pull back against the frame. 90° rotation locks, unlocks fastener.	1. No tools required.	01 min
	"Dzus" type fastener with screwdriver slot. Three-piece, one-quarter–turn fastener. Spring protects against vibration 90° rotation locks, unlocks fastener.	1. Tools may be required. 2. Should not be used for front panel fasteners or in structural applications. Preferred type for lightweight panels other than front panels.	05 min
	Wing head. "Dzus" type. 90° rotation locks, unlocks fastener.	1. No tools required 2. Should not be used for front panel fasteners or in structural applications. Preferred type for lightweight panels other than front panels.	04 min
Knurled .5" extension	Captive fastener with knurled, slotted head. The threaded screw is made captive by a retaining washer.	1. Tools may be required. 2. Operating time depends on number of turns required.	04 min

*When gloves may be worn, increase all dimensions by 0.125 in (0.318 cm).
Each of the fasteners shown in the accompanying table is evaluated from the point of view of ease of use in the context of equipment accessibility for maintenance.

Types of Latches

Type	Description	Maintainability Considerations	Approximate Operating Time
Draw hook latch.	Two-piece, spring latch, base unit and striker. Engagement loop is hooked over striker and lever is depressed, closing unit against force of springs. Lever is raised to unhook.	1. Relatively slow action. 2. Requires considerable force to disengage loops.	03 min
Trigger action latch.	One-piece, bolt latch. Latch is opened by depressing a trigger to release bolt which swings 90° under spring action. To close move bolt back into position.	1. Extremely fast action type latch. 2. Strong spring action might cause personal injury.	01 min
Snapslide latch.	One piece snapslide. Latch is opened by pulling lever back with finger to engage release lever.	1. Fast action.	02 min
Hook latch.	Hook engages knob on striker plate. Handle is pulled up locking in place. To release reverse procedure.	1. Relatively slow action. 2. Takes up room on equipment.	03 min

Types of Fastener Head Styles

Type	Description	Maintainability Considerations
	Fillister head, slotted. Smaller in diameter than round head, but has deeper slot. (FF·S·92 Type 1, Style 4S)*	1. Requires common screwdriver. 2. Deep slot not easily stripped.
	Fillister head, hexagon socket. Same as above except with "Allen" type socket.	1. Required "Allen" wrench not always available. 2. "Allen" wrench usually takes longer to use than common screwdriver.
	Knurled head, slotted. This is an exaggerated fillister with a knurled head and a screwdriver slot.	1. Can be used with fingers or common screwdriver. 2. Preferred type for retaining front panels or removable cover plates.
	Hexagon head. Standard head for machine bolts and screws. (FF-S-92, Type II. Style 10P)*	1. Requires use of wrench or "spintite."
	Hexagon head, slotted. Same as above except with added screwdriver slot. (FF-S-92. Type I. Style 10S)*	1. Wrench, "spintite," or common screwdriver can be used to remove or install.
	82° flathead, cross-recessed. (FF-S-92. Type III. Style 2C)*	1. Cross-recessed screwdriver not always available.
	82° flathead, slotted. (FF-S-92. Type I. Style 2S)*	1. Not suitable on thin panel.
	92° oval head, slotted. Similar to flathead but with rounded head. (FF-S-91. Type I. Style 6S)*	1. Allows deeper slot than flathead. 2. MIL-E-16400 specifies this type with cup washer for rack-mounted panels.
	Pan head, slotted. Large diameter with high outer edges for maximum driving power. (FF-S-91, Type I, Style 9S)*	1. Standard type for panels other than rack-mounted panels and front panels.

*Styles recommended by Federal Specification FF-S-92 for use wherever possible, in order to keep inventories to a minimum.

d = minimum slot depth = 0.0625 in (0.1588 cm).

Types of Captive Screws

Type	Description	Design Considerations
Undercut	Captive screw with undercut. Panel is tapped to screw size.	1. Easily installed and removed.
Spring	Captive screw, spring loaded.	1. Easily installed. 2. Fast action.
	Captive screw which is forced through rubber or plastic grommet.	1. Frequent use can cause excessive wear on rubber or plastic grommet resulting in a loose fit.
Hook	Captive screw with hook which falls into undercut of screw.	1. Hook can be easily jarred loose.
Upset	Captive screw with upset thread.	1. No removal features.
	Captive screw with nut staked in place after assembly.	1. Time consuming to put in place and remove.
Cotter pin	Captive screw with cotter pin forced in place after assembly.	1. Cotter pin can work loose if subjected to excessive vibration.

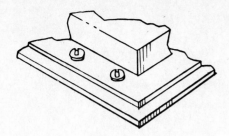

Chassis Mounted on Horizontal Shelf (Secured by Screw Fasteners through Flange)

Securing Method	Maintainability Considerations
	Description: Captive screw. Advantages: No loss of screws and washers. All work performed from one side. Disadvantage: Space required for captive device. Tools Required: Screwdriver. Operating Time: Approximately 6 min per fastener.
	Description: Screw into tapped note with flat washer and lock washer. Advantage: All work performed from one side. Disadvantages: Time required to position washers. Possible loss of screws and washer. Tools Required: Screwdriver. Operating Time: Approximately 8 min per fastener.
	Description: Screw through clearance holes with flat washer, lock washer, and nut. Advantages: No significant advantages. Disadvantages: Possible loss of screws, washers, and nuts. Requires access to both sides of shelf. Two-handed operation. Time required to position washers and nuts. Tools Required: Screwdriver and wrench or "spintite." Operating Time: Approximately 8 min per fastener.
	Description: Screw through clearance holes with lock nut. Advantage: No washers required. Disadvantages: Lock nut difficult to turn. Requires access to both sides of shelf. Two-handed operation. Possible loss of screws and nuts. Time required to position nut. Tools Required: Screwdriver and wrench or "spintite." Operating Time: Approximately 14 min per fastener.

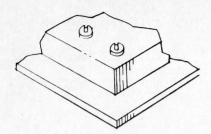

Chassis Mounted on Horizontal Shelf (Secured by Screw Fasteners through Chassis)

Securing Method	Maintainability Considerations
	Description: Captive screw.
	Advantage: No loss of screws or washers. All work performed from one side. No separate handling washers. One-handed operations.
	Disadvantage: Alignment of screw can be difficult. Space required for captive device.
	Tools Required: Screwdriver.
	Operating Time: Approximately 6 min per fastener.
	Description: Stud through chassis with flat washer, lock washer, and nut.
	Advantage: No screw alignment problem. Studs act as locating pins.
	Disadvantages: Possible loss of nuts and washers. Chassis must be lifted over studs.
	Tools Required: Wrench or "spintite."
	Operating Time: Approximately 7 min per fastener.
	Description: Screw into stand-off with flat washer and lock washer.
	Advantage: Very little screw alignment problem.
	Disadvantages: Possible loss of screws and washers. Chassis must be lifted over stand-off.
	Tools Required: Screwdriver.
	Operating Time: Approximately 8 min per fastener.

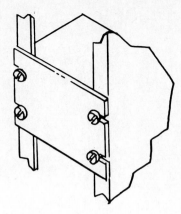

Chassis Mounted on Vertical Rack (Secured by Screw Fasteners into Frame)

Securing Method	Maintainability Considerations

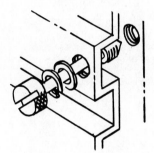

Description:	Captive screw. Slotted head-type preferred for front panels.
Advantages:	No loss of screws and washers. All work performed from one side. No separate handling of washers. One-handed operation.
Disadvantage:	Space required for captive device.
Tools Required:	Screwdriver. None if thumb screw is used.
Operating Time:	Approximately 6 min per fastener.

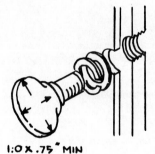

1:0 X .75" MIN

Description:	Thumb screw with lock washer and flat washer.
Advantage:	No tools required. One-sided operation.
Disadvantage:	Possible loss of screws and washers.
Tools Required:	None.
Operating Time:	Approximately 8 min per fastener.

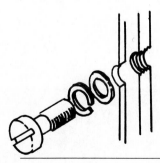

Description:	Screw into tapped hole with flat washer and lock washer. Preferred for panels other than front panels.
Advantage:	All work performed from one side.
Disadvantage:	Time required to position washers. Possible loss of screws and washers.
Tools Required:	Screwdriver.
Operating Time:	Approximately 8 min per fastener.

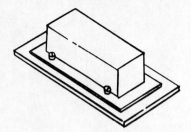
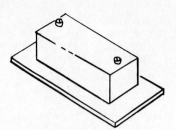

Chassis Mounted on Horizontal Shelf (Secured by Quick-Acting Fasteners)

Securing Method	Maintainability Considerations

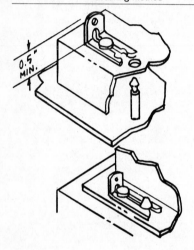

Description:	Snap-slide latch.
Advantages:	Fast action. No tools required.
Disadvantages:	Flange-mounted method may be difficult to operate when side clearance is limited.
Tools Required:	None.
Operating Time:	Approximately 0.02 min per fastener.

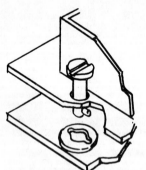

Description:	"Dzus"-type fastener.
Advantages:	Operating time approximately one-tenth that of captive screw-type fastener.
Disadvantage:	Requires use of screwdriver.
Tools Required:	Screwdriver.
Operating Time:	Approximately 0.05 min per fastener.

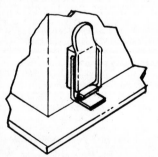

Description:	Spring or drawhook latch.
Advantage:	Requires no tools.
Disadvantages:	Requires space to operate. Closing may require more time due to difficulty in engaging drawhook. Strong spring action might cause personnel injury.
Tools Required:	None.
Operating Time:	Approximately 0.03 min per fastener.

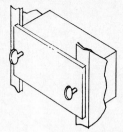

Chassis Mounted on Vertical Rack (Front Panel Secured by Quick-Acting Fasteners)

Securing Method	Maintainability Considerations

Description:	Push-button latch.
Advantages:	Integral handle provides easy removal of chassis. Fast release. Closes with snap action.
Disadvantages:	Requires outside space for handle. Does not pull panel against gasket as well as other types.
Tools Required:	None.
Operating Time:	Approximately 0.03 min per fastener.

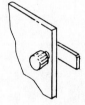

Description:	Pawl latch. Operates with 90° turn.
Advantage:	Requires minimum inside space.
Disadvantage:	Difficult to operate when much pull-up is required.
Tools Required:	None.
Operating Time:	Approximately 0.05 min per fastener.

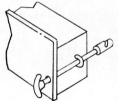

Description:	Cam-action. Operates with 90–180° turn of handle.
Advantage:	Fairly fast release operation.
Disadvantage:	Initial engagement of cam may be difficult.
Tools Required:	None.
Operating Time:	Approximately 0.05 min per fastener.

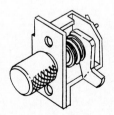

Description:	Adjustable pawl latch. Operating handle pulls pawl toward panel.
Advantage:	No significant advantage over the other types.
Disadvantages:	Requires more time than other types. Excessive inside space.
Tools Required:	None.
Operating Time:	Approximately 0.1 min per fastener.

When frequent removal of a chassis is necessary, the time required for dismounting a hinged cover may be significant. For example, dismounting a chassis using two standard hinges may require the removal of from four to six screws, taking up to 5 min or more. Use of separable hinges can reduce this time to a fraction of a minute.

Any separable hinge must, however, include features to prevent accidental separation under shock or vibration conditions. The hinges shown in the accompanying illustration should not be used to bear the weight of a heavy chassis when the equipment is closed; i.e., guide pins or screw fasteners should be provided to support a heavy chassis.

SEPARABLE HINGES

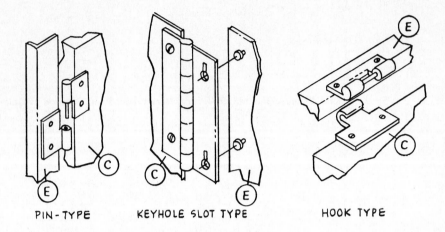

PIN-TYPE   KEYHOLE SLOT TYPE   HOOK TYPE

E = EQUIPMENT

C = CHASSIS

CHASSIS AND DRAWER SLIDE FASTENERS

Slide-mounted equipment chassis systems (and file-drawer slide systems) using ball or roller bearing slides are often improperly designed in terms of the locking and fastening features. Such slide systems should, of course, provide an automatic locking system to prevent a drawer or chassis from being accidentally pulled all the way out of a cabinet. Care should be taken, however, to design the slide-locking system so that it *is possible* to override the lock and fully remove the chassis or drawer. More often than not, this is an extremely complex and confusing operation because the basic procedure is not evident in the hardware itself. Obvious features (i.e., latches or buttons) should be provided on the slide-locking system, and/or clear instructions should appear directly on the hardware. Remember that people seldom remember where they have put an instruction booklet.

More flexible slide-rotational systems may appear even more complex to the user, and thus careful consideration must be given to the release latching concepts and to the visibility and recognition factors associated with locking and unlocking both the horizontal and rotational elements of the slide system mechanisms.

Consider the following guidelines:

The chassis should be easily and quickly detachable for removal from a cabinet without the need for hand tools.

Detaching the chassis from the slide should not require placing the hands in a position that might lead to injury if the chassis suddenly moves.

The slides should automatically lock in the open position and/or in a rotated position securely enough to prevent dislodging the chassis.

The slides should be easily removable from the cabinet for replacement.

Special tools should not be required to remove or repair slides.

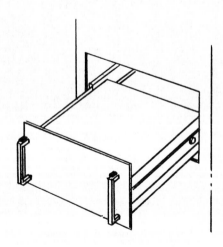

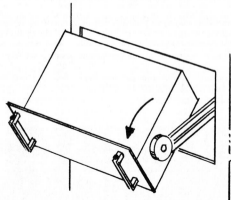

PORTABLE PACKAGING

Human Factors Considerations in the Conceptualization of Various Portable Equipment and Container Packages

To come under the heading "portable," any equipment or container package must be compatible with the limitations of the smallest and weakest military person who may at some time or other have to pick it up and carry it. When a package is too large or heavy to be safely and effectively carried by two people, it should be considered transportable by some other means. The key factors to consider for portable packaging conceptualization include the following.

1. Shape and Size
The shape and size of any portable package must be compatible not only with the user's anatomic characteristics but also with the specific operational intentions, i.e., whether the package is to be lifted and moved only a short distance or whether it is to be carried for a considerable distance, over a flat surface, up or down stairs or ladders, etc.

2. Weight and Balance
The weight of any portable package must, of course, be compatible with the user's anatomic and strength capabilities (not his or her maximum capacity, but some value suitable to the operational requirements). Weight distribution is often a critical variable since a poorly distributed weight may cause the user to lose his or her balance and/or to be unable to manipulate the package for lifting and depositing without dropping it.

3. Grasping and Holding the Package
Consider whether it is desirable to lift and carry the package with one or two hands, to carry it in a "bear-hug" fashion, or to use appropriately positioned handles.

4. Backpack Packages
Design of backpack concepts requires consideration not only of the "fit" of the package as it is being carried (i.e., the proper weight distribution so as not to interfere with walking) but also of the method for picking it up and transferring it to and from the carry position.

5. Contact Hazards
Not only should portable packages be designed so that there are no sharp protrusions, corners, or edges that may cut into hands, but there should also be no external elements that can snag on clothing or catch on things, thus interfering with the safe and orderly manipulation of the package. In addition, there must be sufficient finger clearance along the bottom of the package to preclude mashing fingers when the package is set down.

6. Handholds and Handles
Avoid creating packages without handholds or with handles that are not appropriately placed to aid in properly balancing and manipulating the package. Handles should be designed so that they will not cut into the user's hand or fingers because of the heavy weight of the package. Handle positions may have to be provided for independent handling functions; e.g., the right position for carrying may not be satisfactory for lifting the package or depositing it in a truck or on a shelf.

7. Container Removal
Containers should be designed so that the user removes the container or cover from the equipment or contents it protects, as opposed to removing the equipment or contents from the container.

8. Labels and Instructions
Provide conspicuous and legible labels to avoid handling misuse.

WORKING IN CLOSE QUARTERS

As a general rule, avoid creating situations in which workers have to work in close quarters and/or in the awkward positions illustrated in the accompanying sketches.

When this cannot be avoided, consider the general limiting dimensions provided in these illustrations. Critical factors are the following:

1. There must be sufficient vertical clearance for the worker to enter and leave the environment.
2. There must be sufficient horizontal and lateral clearance for the worker to perform the task (considering the nature, size, and configuration of any tool being used).
3. In the supine position, it is critical that the worker's eyes do not have to be too close to the specific work point; otherwise, the worker's eyes cannot adjust for clear viewing.
4. Openings must be large enough to accommodate both the tool package and the operator's hand, finger, or arms, as shown in the accompanying illustrations.

Note: These opening dimensions are for shirt-sleeve conditions. If heavier clothing may be worn, the openings have to be at least 1 in (2.5 cm) larger on all sides (1.5 in, 3.8 cm, is better).

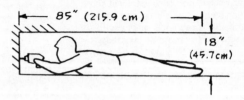

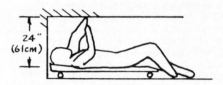

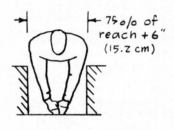

75% of reach + 6" (15.2 cm)

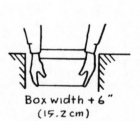

Box width + 6" (15.2 cm)

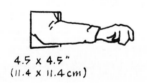

4.5 x 4.5" (11.4 x 11.4 cm)

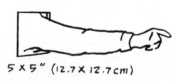

5 x 5" (12.7 x 12.7 cm)

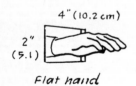

Flat hand

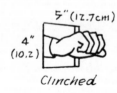

Clinched

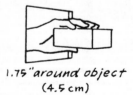

1.75" around object (4.5 cm)

Push button

Two-finger twist (or knob diam + 1.5"; add 1.0" for gloves)

OPERATOR CONTROL PANEL LAYOUT

Basic Principles

Four basic principles should be considered in planning the layout of an operator workplace or, more specifically, the equipment control interface (usually a panel containing controls and displays). These are discussed below.

1. Sequence of Operations

The order in which operators will normally use controls and displays is important in terms of their being able to find what they want and being able to go through the sequence with a minimum possibility of missing a step and also in terms of reducing unnecessary motion due to "backtracking." Although some operations may not have a single sequential pattern, there are usually one or two elements that proceed according to a fairly standard sequence of steps.

2. Functional Grouping

Certain functions within almost any operation are related to each other, and/or certain controls are related to certain displays. Related functions should be arranged together so that their association is readily apparent to the operator. When such functions are not grouped, operators tend to get lost and to wander from one point to another trying to find what they want. Although there may be a few elements within the workplace that cannot be placed together for obvious reasons (i.e., a foot pedal obviously cannot be placed where an associated visual display would be), an attempt should be made to avoid scattering controls and displays anywhere on the panel without regard to functional associations.

3. Frequency of Use

Some controls or displays will probably be used more frequently than others. These should be located so that they are maximally convenient; i.e., visual displays should be near the normal sight line, and controls should be near the hand when the arm is in a normal position. For example, a frequently used visual display should not be located where the operator must continually turn his or her head or body in order to see it, and a control should not be placed where the operator continually has to reach a long distance for it.

4. Emergency Use

Once in awhile there will be a control that is critical under emergency situations; i.e., it may be vital that the operator get to it and operate it successfully in the shortest possible time. Obviously, if such a control is hidden, out of reach, or otherwise difficult to get to and operate, the operator may not be able to reach it in time.

Note: Each of the above principles must be considered in light of a particular operator interface problem; there is no set priority that can be all-prevailing.

Positioning Panel Elements Relative to Line of Sight

Visual displays, including indicator lights, instruments, panel and control labels, etc., are useless to the operator if he or she cannot see them. This requires that no control (joy stick, steering wheel, etc.) be placed between the observer's eye and other displays. In addition, it requires that care be taken in the placement of control and display labels so that when critical, the operator's hand will not cover up an important label that may be used in adjusting the control.

As a general principle, labels should be located in a consistent manner with respect to being above or below a control or display; i.e., avoid having some labels above a control, while others are below a control—at least on the same panel.

However, the choice of label position should also reflect consideration of the expected operator's eye reference; i.e., if the control is above the eye reference position, the label is more likely to be visible than if it were placed above the control.

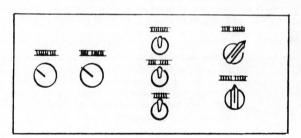

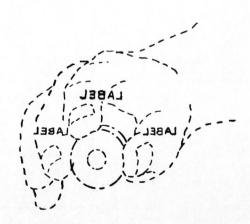

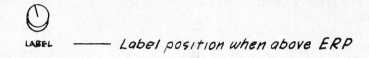

LABEL —— Label position when above ERP

left of ERP —— LABEL

LABEL —— right of ERP

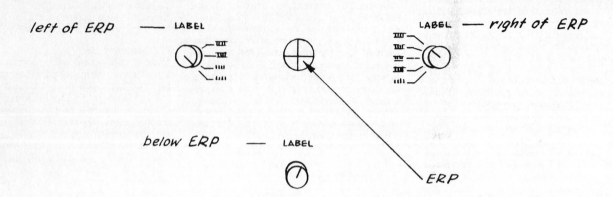

ERP

below ERP — LABEL

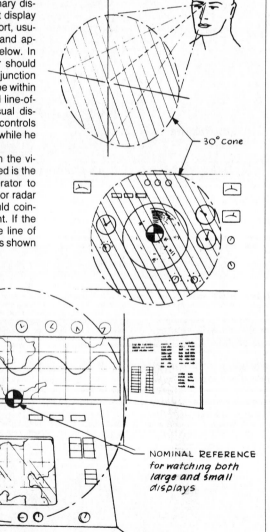

Primary Visual Reference Principle

When it has been determined that an operator is expected to visually monitor a primary display for a great share of the time, that display should be located for maximum comfort, usually directly in front of the operator and approximately at eye level or slightly below. In addition, the things that the operator should be able to see fairly regularly in conjunction with the principal viewing task should be within about a 30° cone around the principal line-of-sight line. This includes not only visual displays and indicators but also critical controls that the operator may have to look at while he or she is adjusting them.

A typical console situation in which the visual reference principle must be applied is the large CRT display requiring the operator to continually watch for and track sonar or radar targets. The center of the CRT should coincide with the comfortable line of sight. If the operator has to share more than one line of sight, seek a compromise reference, as shown in the accompanying illustration.

30° Cone

NOMINAL REFERENCE
for watching both
large and small
displays

Display Arrangement

When the circumstance includes an array of displays that must be monitored on a regular basis to determine whether conditions are approximately "normal," the check-reading task is made easier for the operator if the displays are arranged so that their indications (e.g., pointer positions) generally reside in the same relative place on the instrument (see the accompanying sketch). Arrange the normal pointer reading at the twelve o'clock position when the array is vertical and at the nine o'clock or three o'clock position when the array is horizontal.

If displays are ordinarily read in sequence, arrange them in order, either horizontally or vertically. The normal reading pattern is from left to right or from top to bottom.

When a display is adjustable by means of an associated control, the control should be located as close to the display as practical. However, since there may be a directional relationship, as shown in the accompanying sketch, care must be taken to position the control relative to the display so that there is no doubt in the operator's mind as to the direction in which the control should be moved in order to cause the displayed element to move in the right direction. Controls which are placed below their respective displays are less confusing than those which are placed to the right or left of the displays.

When it is impractical to place controls close to the displays they affect, try to arrange both the displays and the controls in similar patterns (see the accompanying sketch) so that there is a similarity between the two.

When a control causes something to move in space outside the console (e.g., an antenna), control-element directions of motion should also be similar, i.e., clockwise, counterclockwise, left to right, up and down, etc.

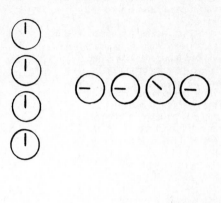

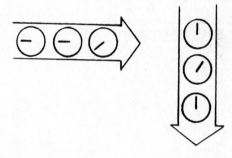

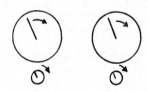

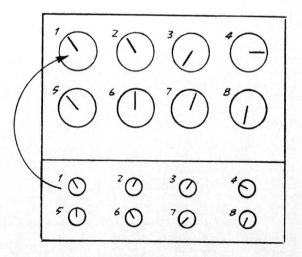

An important objective in panel layout should be to make the functional grouping of controls and displays readily apparent to the operator. There are various effective methods for doing this, some of which are illustrated in the accompanying sketches:

1. Spatial separation is effective when there is sufficient panel area. This is the preferred method.
2. When the panel is necessarily space-limited, one can use lines or borders around the various functions. This is particularly effective if the functional grouping is irregular.
3. If one functional group is especially important for the operator, it can be made more conspicuous by using a different color for the entire background for the group from that which is used for the rest of the panel.
4. Subpanels—either the protruding or the inset type, as shown by the accompanying sketches—are effective for setting off the various functional groupings.
5. In some cases, a separate package can be attached to the main package, especially if the package is small but needs a special physical orientation to make it easier to use.

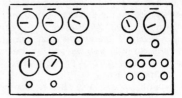

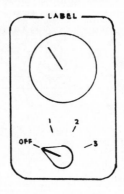

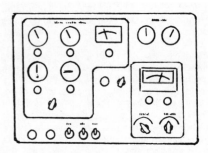

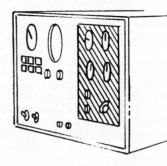

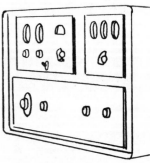

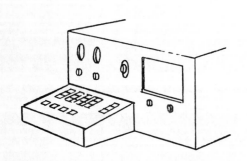

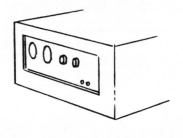

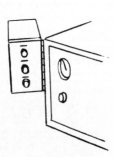

CONTROL LABELING

As a general guideline, all controls should be labeled. Exceptions, of course, are control devices that everyone recognizes and knows how to operate and controls that are located where the operator typically cannot see them. Examples of these exceptions include the following:

Automobile steering wheels and aircraft joy sticks
Automobile foot pedals and aircraft rudder pedals
Door and equipment cover latches
Switch mats, etc.

Rotating controls should not be labeled on the control knob or handle because the label becomes difficult to read when the knob is rotated.

When labels associated with controls are placed on the panel adjacent to the control, position the label so that the control knob will not obscure the label from the expected position from which the observer looks at the control. That is, when the eye reference is below the control, place the label below the knob; when the eye reference is above the control, place the label above the control; and when the eye reference is to one side of the control, place the label on the observer's side.

Otherwise, try to be consistent in positioning control labels; i.e., minimize the alternating positioning of labels above and below on the same operating panel.

When a control may be operated in the dark, provide illumination for the labels. This applies to all control labels, not just those which the designer may have decided were important. The reason for this is that operators make their own decisions relative to importance, and an unlabeled control may be inadvertently operated and/or confuse an operator because he or she assumes that the illumination is malfunctioning.

Refer to the section on visual displays for other recommendations regarding labeling.

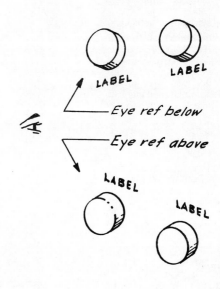

STORAGE SPACE AS A SPECIAL WORKPLACE

The location and design of a storage space should receive the same systematic appraisal and attention as other work stations, especially when the storage space is utilized frequently. Above all, avoid the temptation to treat the problem of storage space as something to be taken care of after all the more important operator work-station problems are solved. Design of the storage subsystem should consist of more than "making do" with whatever leftover space one can find in the facility or vehicle.

1. General Principles
Provide storage space where it is needed, not at some distant, out-of-the-way, or inaccessible location.

Design the space to "fit" the items to be stored; do not design just a generalized volume of space or shelf area.

Provide enough space not only for the predicted use but also for unexpected or future expansion.

2. Content Considerations
Size, shape, and weight

Special storing conditions, e.g., package position and protective and/or safety precautions to prevent deterioration and damage

Label exposure, i.e., the necessity to have identification labels immediately accessible as the storage container or storeroom door is opened and the contents are viewed

3. Special Feature Considerations
Content separation, securing, and preserving

Content adjustment as needs change

Ease of ingress and of storage and retrieval

Ease of cleaning the storage area (dirt and spills)

Content security, i.e., inaccessibility to unauthorized persons

Illumination so that the user can move about safely and/or can see to identify and manipulate stored items, make space adjustments, or clean

Environmental control, i.e., control of heating, cooling, ventilation, humidity, shock and vibration, etc.

4. Key Element Design Considerations
Location

Shelving and partitioning

Doors and covers

Handles, latches, and locks

Content securing

5. Storage Type and Location in Relation to Order and Frequency of Use
In many situations storage cabinets are built with little thought given to what type of storage is needed at what locations, e.g., a kitchen in the typical home. Although this is not always the case, kitchen cabinets may be created merely to fill in the spaces between appli-

ances. The cabinetmaker may be told to provide a variety of drawers and cabinets to fill specified dimensions. As a result, the user often finds that things cannot be stored next to the area in which they are needed because the storage space provided does not accommodate them. Examine the use sequence and/or the frequency for each given storage requirement and make sure that the storage plan has some direct relationship to the storage and retrieval needs of the eventual user.

Storage Shelf Depth in Relation to Shelf Height and Package Weight

Avoid the temptation to provide general shelf storage space without regard to human limitations in terms of reaching and lifting. Obviously, people cannot reach very far into a shelf that is very high or very low. In addition, people cannot support or manipulate as heavy a package when a shelf is either very high or very low as they can when the shelf is somewhere between waist and shoulder height.

The accompanying sketch provides some general guidelines with respect to package weight and shelf height. Obviously, these are not absolute values, since the size, shape, and other geometric features of specific packages tend to modify the capabilities of the individual who is storing or retrieving them. Note that, because of a balancing problem, a worker in the kneeling position cannot support a very heavy or large package.

The accompanying sketches illustrate the difficulties created by extremely deep shelving. The following are general guidelines for specifying shelf depth:

1. Below waist height, the maximum shelf depth should not exceed about 18 in (46 cm).
2. Above shoulder height, the maximum shelf depth should not exceed about 12 in (31 cm).
3. At waist to shoulder height, the maximum shelf depth should be about 24 in (61 cm).

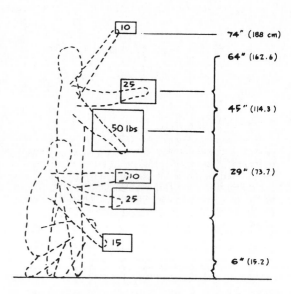

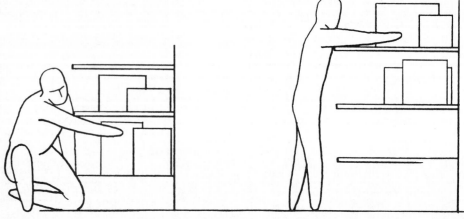

Support Furnishings

GENERAL GUIDELINES FOR FURNITURE

In designing or selecting furnishings, the following general considerations should be kept in mind:

User fit: Careful consideration should be given to the dimensions and anatomic characteristics of users to make sure that the furniture fits them, supports them properly, and adjusts to their activities, i.e., rest, work, and recreation.

User efficiency: Consideration should be given to what users have to do to and with the furniture in terms of arrangement of elements, articulation of components, operation of special controls, accessibility within components, and visual impact of finishes.

Interactive characteristics: Consideration should be given to how one piece of furniture interacts with another positionally and in terms of orientation and interference.

Safety: Consideration must always be given to potential safety hazards, including sharp contact points, fragility of the structure, balance and stability, and flammability.

Housekeeping: Someone has to clean around, move, store, and rearrange furniture at some time. The design should aid, not hinder, such operations.

Good furniture design can be created for any styling objective, but good human engineering must come first.

The accompanying illustrations have been made purposely simple to establish the idea that the basic furniture configurational requirements generally have little to do with aesthetic flourishes. The designer, using his or her imagination, can add the styling flourishes around any basic structural concept.

DESIGN CHARACTERISTICS OF A GOOD SEAT

Whether designing a chair, a davenport, or a pilot or vehicle driver seat, one should start with the following critical features as a base line:

Good posture: To provide good sitting posture, i.e., one in which the back and neck muscles are least strained, the seat pan should tilt back approximately 5° to shift the upper torso weight and thus cause the torso to rest fully against the seat back. The angle between the seat pan and the seat backrest must be approximately 105° to keep the torso against the backrest and yet not force the occupant to lean his or her head forward in order to balance it properly.

Surface support: The seat should be approximately 19 in (48.3 cm) wide to keep large occupants from "lapping over." The seat-pan length should be about 17 in (43.2 cm). A longer seat strikes the back of the short person's legs; a shorter one provides too little support for the longer-legged person's thighs. The seat back should be about 20 in (50.8 cm) high so that even the tallest person's shoulder blades are supported.

Seat-pan height: The best compromise seat-pan height for a full range of male and female adults for normal sitting (i.e., chairs) is about 17 in (43.2 cm) at the leading edge of the seat pan. If the seat pan is higher, the short female occupant's legs will dangle. Even the 17 in (43.2 cm) assumes that she will be wearing shoes with a 1.5- to 2.0-in (3.8- to 5.1-cm) heel. Although a slightly lower seat pan would be desirable for very short-legged people, the lower height would make it very difficult for tall people to get up out of the seat.

For other types of seats, all the above guidelines, except for the seat height, should be used.

Stools can be higher, but a footrest should always be provided to maintain the basic foot-to-seat-height relationship.

Vehicle seats should generally be lower, at least for the driver, because the driver needs to extend his or her legs to operate the pedals. The optimum height for an automobile driver's seat should vary from about 12 to 13 in (30 to 33 cm) above the nominal floor (heel resting position for accelerator pedal operation). In order to accommodate the range of leg reaches from that of the 5th-percentile female driver to that of the 95th-percentile male driver, the seat should have a minimum of 8 in (20 cm) of fore-aft adjustment; i.e., when the seat is all the way back, the seat height should be 12 in (30 cm) above the heel point, and when it is all the way forward, the seat height should be 13 in (33 cm). This seat-height variation, coupled with the fore-aft adjustment, allows the short driver to see over the hood and still reach the foot pedals, and the lower aft position allows the taller driver to sit low enough to clear his or her head.

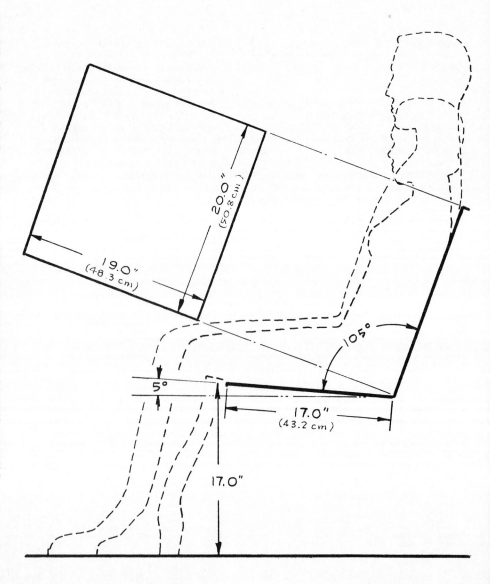

Seat Design: Cautions and Suggestions

Except in unique cases (such as the astronaut couch designed to help absorb launch and reentry *g* forces), avoid the use of seat-pan or backrest contouring. Sling-type seats are especially bad. Contouring might work if every seat were designed specifically for each occupant, but this is not practical. No two people need contours in the same place, and a contour in the wrong place is much worse than none at all. Fortunately, most people are appropriately padded so that they can accommodate well to a flat seat pan or seat backrest.

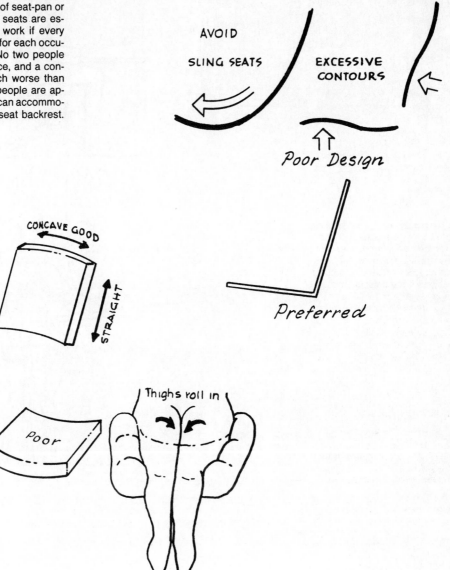

Deep cushions and/or softly sprung cushions should be avoided for both the seat pan and the backrest. As the accompanying sketch illustrates, the upper part of the backrest pushes the occupant's torso forward, his or her own weight compresses the lower back and seat cushions to reduce the desired included angle, and the forward cushion cuts off circulation in the legs.

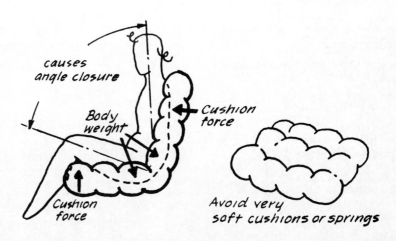

OPERATOR CHAIRS Provide personal height adjustment for both the seat and the backrest (a most important feature for the secretary). The backrest should also tilt so that it can conform to the individual worker's lumbar support needs.

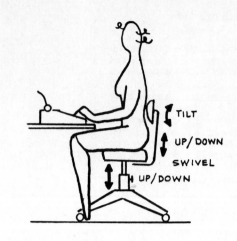

WORKBENCH STOOLS Provide essentially the same adjustments for personalization that were suggested for the operator chair, except do not provide castors; i.e., this seat must "sit still." Make sure that the vertical adjustment systems remain secure if the chair is lifted by the backrest.

Note: Some experimental stools have been suggested which have tilting seats (i.e., for a proposed "half-sit, half-lean" position). There is no evidence that such a device improves a worker's productivity.

The operator chair should have dimensions and seat and backrest angles similar to those shown in the accompanying illustration.

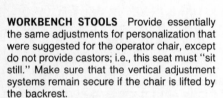

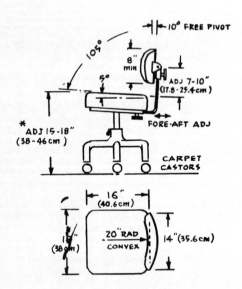

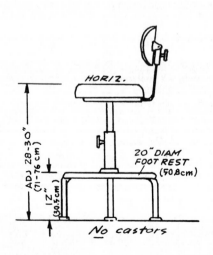

Domestic Products

GENERAL

Human Factors in Consumer Product Conceptualization and Design

There are at least three distinct, but not necessarily mutually exclusive, areas in which human factors should be considered relative to the conceptualization and design of consumer products such as appliances, entertainment and recreational equipment, tools, and toys. These are discussed briefly below.

1. Safety

Products should not be designed so that they could fail and cause injury to the user or so that they could easily be misused in a manner that could lead to injury to the user, nor should they be made out of materials that might be toxic, poisonous, or otherwise hazardous to the health of the user. Lack of a proper safety analysis during the conceptual design stage typically results in the following:

a. Products in which toxic or flammable materials are used
b. Products which are made of breakable materials or which have sharp protrusions, corners, or edges
c. Products in which exposed moving parts cause pinches, cuts, or amputations
d. Products that can be used improperly to strike another person or to shatter glass
e. Products that can be swallowed
f. Products that can cause electric shock because of improper insulation or grounding
g. Products that people can easily fall from
h. Products that can produce injurious noise, extreme heat, or flying particles that could puncture a person's skin or eye
i. Products that could cause rupture or strain when lifted or injury to a foot if dropped
j. Products that can cause burns

2. Operability and Maintainability

In order for a product to be truly acceptable, it has to be easy to operate and maintain. Most consumers fail to appreciate this aspect of a product fully until after they have purchased it. There is considerable evidence that the typical consumer will not purchase another product of the same make once he or she has discovered that it is difficult to operate or maintain. Typical features that show up in many consumer product designs are the following:

a. The product is hard to pick up, hold onto, manipulate, or maneuver.
b. The product is hard to open and close, and the latches and handles are difficult to manipulate.
c. The controls and displays are poorly labeled, hard to read, difficult to understand, and awkward to adjust or operate.
d. The product is difficult to load and unload and requires the user to stretch, reach, bend, etc.
e. The product requires too much force or too much precision to operate properly, resulting in frequent errors in use.
f. The product is packaged in such a way that it is difficult to get to parts that need adjustment or replacement.
g. A presumably movable product is not provided with appropriate handholds or casters so that it is easy to move to a new location.
h. The product contains hard-to-find parts, with the result that acquiring replacement parts later is impossible or time-consuming.
i. It is practically impossible to keep the product properly adjusted long enough for it to do its job.
j. The product wears the operator out because of its weight, the awkward way it has to be held, the noise it makes, or the way it vibrates.

3. Attractiveness

Although most manufacturers and designers are fully aware of the importance of product attractiveness to salability, and thus pay more attention to this aspect of product design than to any other, attractiveness is an elusive variable when one tries to define it. A critical comment is in order also: Although attractiveness is an admirable and desirable trait in product conceptualization, it is extremely important that, in an effort to make something attractive, the designer be careful not to compromise product safety, operability, or ease of maintenance.

Since attractiveness is truly a human factors consideration, it is important to recognize that it may be a transient phenomenon; i.e., what one considers attractive this year is not necessarily considered attractive several years from now. This point is made to refute the argument that the aesthetic function is the most important consideration in consumer product conceptualization; i.e., a product that works well can continue to do so over many years and through many different packaging treatments, but only if the initial conceptual objective was to make the product work well. It should also be emphasized that a product can be made attractive after the objectives of operability and maintainability have been developed. Some of the hazards of starting with the aesthetic objective are the following:

a. Aesthetic features may interfere with product recognition. For example, there is often a tendency to camouflage a product, i.e., to make it look like something that it is not.
b. Aesthetic features may make visually displayed information impossible to read, as when white labels are printed on bright gold metallic backgrounds.
c. Exotic-looking control knobs may interfere with the user's ability to adjust the control easily and precisely, or the fancy surface treatment or shape may actually hurt the user's hand or fingers.
d. A streamlined cover or case may make it difficult to pick up the device or may cause a person to drop the object because there is no way to get a good grip on it.
e. An attempt to completely hide all visible

125

means of fastening a cover on a product may cause the owner to damage the cover while trying to get it off for maintenance or replacement of a part.

f. Ornaments placed on a product to make it attractive often turn out to be hazards, or at least annoyances, because the user is constantly catching his or her clothing or hands on the ornamentation.

g. Color used for aesthetic reasons often becomes "visual noise" around an important warning display, making the display insufficiently conspicuous to attract the user's attention.

h. Fancy logos are often much more conspicuous, readable, and intelligible than more important user instructions. In some cases, important labeling is actually left off because the designer felt the label would disturb the overall aesthetic appearance that he or she was trying to create.

i. Furniture and related products are often less than useful, operable, comfortable, or movable because of the interferences created by aesthetic objectives. For example, deep, puffy cushions do not provide needed posture control, and they make it impossible for the short-legged user to sit comfortably.

j. Products in which sharp edges and corners are typical styling features often injure the user.

Misuse Behavior as a Key Safety Consideration

Liability should be an important consideration in conceptualizing any consumer product; i.e., the possibility of consumer injury and a potential lawsuit should stimulate any planner to investigate thoroughly the possible failure modes of a proposed product design, including especially the misuse behaviors of expected consumers.

Avoid the excuse that "I can't worry about every moron who may buy or use my product." These words have come home to haunt many manufacturers and, more recently, many designers. Worrying about potential misuse of a product can become an advantage in that it generates product features that will make the product more salable, e.g., easier to operate without making mistakes.

The following are some of the typical misuse behaviors that consumer product planners should consider.

1. Children

a. Children are extremely curious. As a result, they investigate and examine or try to play with products or product elements which they should not touch. If there is a potential hazard, design the product so that children cannot reach or operate the device.

b. Small children investigate by touching things, putting their fingers or hands into openings, putting things into their mouths, or at least trying to put their tongues on things. Make sure that objects that are small enough to swallow cannot be removed, that holes are either too small to allow passage of fingers or arms or large enough so that fingers and arms can be pulled back out, and that surface finishes are not toxic and cannot flake off.

c. Children like to climb or reach for something on top of a surface which they cannot see. Package products so that there are no climbing aids. Design hazardous products, like stoves, so that children cannot reach the controls or burners.

d. In spite of a designer's best intentions, most children will use a toy in an entirely different manner from the way it was intended to be used; e.g., they will use it to pound with, stick it in their own or another child's eyes, or try to pull it apart to see what makes it work. Make sure that toys are sturdy enough so that they will not break or come apart; soft enough so that they will not break something else or injure another child; devoid of sharp edges, corners, and projections; and made of materials that are not toxic.

e. Children are unsteady on their feet, and they often run so fast that they are unable to maintain their balance or control their bodies. As a result, they often fall or run into objects. Products around which children may play should be designed with the idea that a child might fall or run into them; i.e., there should be no sharp edges or corners, and they should be made of unbreakable materials.

2. Adults

a. Adults often attempt to operate a product without looking at the instructions first. As a result, they may do something to create a hazard for themselves and/or for others around them. Products should be designed as nearly as possible to be understandable without requiring the user to read special instructions; i.e., the critical instructions and labels should tell how to start, adjust, and turn off the product.

b. Adults are often absentminded or easily distracted, and they often forget to be careful. Products that have a hazard potential should have safety lock-out features and/or added protective features that will minimize the hazard in the event the user forgets and turns a wrong knob, for example.

c. Adults may be in a hurry, may bump into things in their haste, or may not look to see where they are putting their hands or feet. Products should be designed so that even if the user does tend to hurry, injury will not result.

d. Adults tend to believe that they are smart enough to take shortcuts (even though there may be a warning to the contrary). Often these shortcuts end in tragedy. Products that have such hazard potential should be designed so that shortcuts are out of the question.

e. Many adults are willing to take risks, even though they know there is a chance of making a mistake or a miscalculation, either because they think they can get away with it or because they truly believe their time is more important. Products should be designed so that they cannot be operated in any way other than the right way; i.e., there should be no opportunity for risk taking.

3. The Handicapped

Many handicapped consumers are forced to try to operate products that were not designed with their infirmities in mind. As a result, they are subject to hazards that other adults are not exposed to. Since the number of handicapped persons is increasing and since they are becoming a significant portion of the consumer product market, products should be designed so that people with handicaps can use them efficiently.

Design Conceptualization for an International Consumer Market

Designing a product that may be marketed around the world requires knowing something about the differences between consumers that are critical to the product's safe and efficient use and to its acceptability. The following are some of the human factors that are important to an understanding of a product's operation and acceptability in different cultures.

1. Anthropometric Differences

People of various nationalities differ in terms of their body dimensions. Some are taller, heavier, and stronger than others. Some have larger hands, longer reaches, and more strength than others. Products designed for American consumers might be too large or too high or have reach requirements that are too great for Oriental consumers. By the same token, certain products designed for the smaller Oriental might not provide sufficient clearance for the average American.

2. Technological Experience

An ethnic population that is not used to many of the modern products that a more technically oriented population is used to will neither appreciate the technical complexities that may be built into a product nor understand how to operate or maintain the product. Thus the complexity of a product should be geared to the expected user population. Even though a product may not seem as sophisticated and efficient as it could be, its final effectiveness and acceptance might be much higher if it were less complex.

3. Language Barriers

A major problem for designers of equipment that may be used in many parts of the world is the difficulty of labeling a product and/or creating appropriate instructions for each potential user. Translating instruction manuals is fairly simple; however, labeling to provide universal understanding presents a more difficult problem. As will be noted elsewhere in this handbook, using international symbols in labeling is one way to ensure that people who speak different languages can still understand and operate a product or device.

4. Ethnic Differences in Terms of Styling

Although of less importance than in past years, various styling features have considerable significance to members of some religious or ethnic groups. It should be noted that much of this can be treated in a cosmetic manner; i.e., after a basic product package has been developed for the total market, cosmetic features appropriate to particular user markets can be added to increase user acceptance.

PACKAGING

Specific Human Factors Considerations in Consumer Product Packaging Conceptualization

1. The package should be easily movable and transportable.
2. The package should be designed so that it will "stand on its own"; i.e., avoid a design that has to be propped up or held in place.
3. If the device has to be assembled before it can be used, design the components so that the parts are easily identifiable, so that it is obvious how they fit together, and so that they will go together without forcing and without using special tools.
4. If the device has to be disassembled between uses, design the parts so that special tools are not required to take it apart.
5. Use captive fasteners so that they will not be lost.
6. If assembly or disassembly instructions or other warnings or instructions are critical to the preparation, operation, or maintenance of the product, place these instructions or warnings on the device at points where the user normally would see them as he or she begins assembly or operation.
7. The package should not have any spring-loaded devices that could pop out when the product is disassembled.
8. The package should not have any parts which are sharp or pointed or which are so flimsy that they are easily bent or broken. If sharp features are necessary to operation of the product, provide pro-

tective covers so that the user is not accidentally stabbed or cut.

9. The product should be packaged in such a way that it is easy to clean; i.e., avoid narrow slots or crevices in which dirt could become lodged.
10. The package should have smooth surfaces that will not snag hands, clothing, cleaning cloths, etc. Use exterior materials and/or finishes that are minimally subject to wear and damage.
11. The package should be shaped and/or have appropriate handholds so that the user can manipulate or hold the package in place. Provide adequate finger clearance so that the package can be set down without pinching the user's fingers.
12. If necessary, the package should be designed so that it can be stacked and/or easily placed in a carrying case or shipping carton.
13. The package should be designed so that internal parts cannot be shaken loose while the product is being moved.
14. The package should be designed so that liquids, powders, or similar materials are not easily spilled when the product is moved.

Human Factors Considerations in Packaging for Attractive Appearance

There is an almost infinite variety of possibilities with regard to making products attractive, which is an important human factor in encouraging the consumer to purchase a particular product. However, it is extremely important that the appearance factor not interfere with the operability, maintainability, or safety of the product.

The principal areas to consider when adding to the attractiveness of a product are the following:

1. Color
2. Pattern
3. Surface texture
4. Shape features

The following guidelines concern pitfalls that should be avoided when adding to the attractiveness of a product:

1. The product should look like what it is. Avoid the temptation to create a concept that camouflages the real identity of the product. Although sometimes there is an aesthetic value to making a product blend with a general decorative scheme, this usually can be done without destroying key product identifying features.
2. The general appearance of the product package should provide some clues as to what the observer is supposed to do with it. Once again, avoid the temptation to camouflage the features that are obvious indicators of how the product is to be approached and used. Typical er-

rors include a camouflaged latch, handle, control device, door, or drawer; the purposeful omission of a label; the minimization of visual contrast between labels and backgrounds; and the stylization of a label to the point where the observer cannot tell whether it is a label or merely a decorative feature.

3. The product's appearance should imply simplicity, not complexity. The typical consumer has come to expect apparent organization among the various exterior product features, the absence of extraneous features that he or she obviously will not use, and a general product appearance that is uncluttered by bizarre patterns that tend to hide the key user interface elements.
4. The product should imply efficiency. The first impression should dispel any feeling that the product is crude, cumbersome, out of balance, etc. For example, the typical consumer has been conditioned to associate solid structure with sturdiness and strength, symmetry with balance, and smooth lines with efficiency. Although an asymmetrical package is quite acceptable as long as the observer recognizes that the asymmetry is purposeful, he or she will be disturbed if the lack of symmetry looks like an oversight. Do not confuse "streamlining" with efficiency (a typical error in the design of certain products such as chairs, tools, and hand-held appliances).
5. The product should appear coherent and complete to the observer. Avoid the use of features that create the impression that the product was thrown together, that parts were merely "tacked on," or that something might be missing. Avoid features that may be difficult to assemble correctly; otherwise, it may appear as though the various elements of the product do not fit properly. Many packages have failed to reach their appearance objective because there was too much opportunity for the factory assembler to attach parts in such a way that they looked misaligned.
6. The product's proportions should relate to the human scale. The product should not overwhelm the user psychologically, and the parts should not be so large or so small that the user has difficulty getting hold of or manipulating them.
7. The shape of the direct interface devices (handles, knobs, controls, etc.) should be compatible with what the user has to do with them. Avoid the temptation to overdecorate such interfacing devices so that they are hard to get hold of or so that they may hurt the hand when they are gripped.
8. The visual contrast required for specific visual display visibility should not be

compromised by colors or patterns used to decorate the exterior of the package.

9. Appearance constancy should be addressed in terms of making sure that the product is recognizable and attractive from all angles from which it will be viewed. Avoid the tendency to concentrate on only one side, which gives the impression that the designer did not care about the other sides.

10. Avoid package features that create odd illusions, e.g., a pattern of lines or other geometric features that look like a grinning face or tilted stripes that make the package look as if it is leaning.

11. Choose colors which provide "life" and visual interest but which are not so intense that they are overwhelming or conflict with the color environment within which they will appear. Do not use colors for aesthetic purposes that will conflict with specific operational colors; i.e., if color codes are used, make sure that these codes are not rendered ineffective by the colors used as a background.

It is easy to create undesirable illusions when implementing certain appearance objectives. Beware of the following:

1. A package that appears larger at the top than at the bottom gives the impression of instability or top-heaviness.
2. A package that appears too tall and slender gives the impression that it may fall over.
3. Evenly spaced concentric circles or spirals appear to alternately advance and recede, thus creating visual pulsation.
4. An extended straight line or edge appears to sag in the middle.
5. Crossing lines appear to bend near the point of crossing.

On the other hand, desirable illusions can be created, such as the following:

1. A positively sloping line can give the illusion of forward movement, and a negatively sloping line can give the illusion of slowing down; e.g., vertical patterns that slope toward the direction in which a vehicle would ordinarily move can make the vehicle look as though it is moving forward, even though it is standing still.
2. A long horizontal line along the side of a short automobile will make the car look longer.
3. A very tall, narrow package or wall panel will be less disproportionate if a horizontal line crosses the expanse about midway.
4. A border around the edge of a package form provides the observer with a firmer basis for judging the package shape and size.
5. The overall size of a package is more pleasing to the observer if it is compatible with the sizes of other packages with which it will be viewed. This suggests that, in some cases, a product should be packaged in a larger container than is actually required to house the internal components.
6. Consumers frequently associate bulk

(e.g., an overstuffed chair) with comfort. Even though the bulk may not be necessary, the illusion it creates might be necessary in order to make the user feel that the product possesses what he or she considers to be desirable features.

PRODUCT SAFETY

Safety Concepts by Product Category

Although the following safety considerations by no means represent all the potential safety problems to be addressed for each of the product categories, they provide a point of departure for planning product design concepts.

1. **Electrical Appliances**
   a. Electrical grounding
   b. Overload fuses
   c. Nonmetallic user interface hardware
   d. Automatic power cutoff with cover removal
   e. Adequate insulation
   f. Warning and caution labels and instructions for operation and maintenance

2. **High-Temperature-Producing Products**
   a. No direct contact between the internal heat source and the case
   b. Insulation to attenuate radiant heat transfer to the exterior
   c. Low-heat-transfer hardware to user interfaces (e.g., handles and controls)
   d. Nonflammable materials and finishes

3. **Mechanical Products with Moving Parts**
   a. Covers over moving gears, chains, belts, and activating levers
   b. Captive spring-loaded parts
   c. Fracture-, chip-, and distortion-resistant metal, composition, or plastic parts

4. **Products with Explosive Components**
   a. Nontoxic substances
   b. Minimal energy release commensurate with purpose
   c. Structural integrity of the container
   d. A replacement package and a method of replacement that preclude mishandling and other forms of misuse

5. **Toys for Small Children**
   a. Nonflammable materials
   b. Nontoxic materials
   c. Unbreakable materials
   d. Sufficient size to preclude swallowing the toy or putting it far enough into the mouth to strangle and no small parts that can be removed and put into the mouth
   e. No sharp corners, edges, or protrusions
   f. No cords or wires that could be swallowed or wrapped around the user or a playmate
   g. No metal parts that could be inserted into an electric outlet
   h. No articulating element in which the child could be pinched

6. **Tools and Cutlery**
   a. Appropriate electrical and mechanical component grounding and guards
   b. Guards to prevent contact by flying particles or sparks

   c. Safety catches and covers
   d. Heat insulation around heat-producing parts (nonflammable)
   e. No cords, if possible; otherwise, quick release connectors
   f. Appropriate warning labels and instructions on the product
   g. Safety features associated with replacement of parts and cleaning or servicing

7. **Furniture**
   a. Structural integrity for expected loads
   b. Antitip construction
   c. Nonflammable and nontoxic materials
   d. No sharp corners or edges
   e. Lockable casters (when used)
   f. A foldable design that will not cut or pinch the user or collapse accidentally during use and a design that lets the user know when the product is fully locked in place

8. **Other Products**
   a. Lockable medicine cabinet doors
   b. Refrigerator doors that can be opened from inside in the event a child gets shut in accidentally
   c. Stove burners and controls arranged so that a child cannot get to them
   d. Space heaters designed so that children cannot be burned by hot exteriors
   e. Ladders which will not collapse and which the user can be sure are locked in place
   f. Stair rails designed so that small children cannot get their heads caught between the rail supports or fall between the supports
   g. Baby furnishings designed to prevent the child from falling between the mattress and the rails, from lowering a rail or getting caught between the rail supports, and from being cut or pinched by rail-lowering hardware; nontoxic materials and finishes on parts of the product that a child might touch with his or her mouth

In conceptualizing any consumer product, consider "Murphy's law": If anything can possibly go wrong, it will.

SELECTED READING LIST

DeGreene, Kenyon B.: *Systems Psychology*, McGraw-Hill Book Company, New York, 1970.

Gagne, Robert M. (ed.): *Psychological Principles in System Development*, Holt, Rinehart and Winston, Inc., New York, 1962.

Hammer, Willie: *Handbook of System and Product Safety*, Prentice-Hall, Inc., Englewood Cliffs, N.J., 1972.

Harrigan, John E., and Janet R. Harrigan: *Human Factors Program for Architects, Interior Designers and Clients*, Blake Printing and Publishing, 1976.

Meister, David: *Behavioral Foundations of System Development*, John Wiley & Sons, Inc., New York, 1976.

Meister, David, and Gerald F. Rabideau: *Human Factors Evaluation in System Development*, John Wiley & Sons, Inc., New York, 1965.

Work Environment

NOISE

Equipment should be designed so that it does not contribute materially to the noise environment of a workspace—first, to make sure that it does not interfere with necessary verbal communication, and, secondly, so that it does not introduce irritating, stressful sounds that may contribute to operator fatigue.

TEMPERATURE

Although individual equipment elements generally do not seriously influence the thermal environment, it is important to select a workplace area for electronic-computer equipment and for workstations keeping in mind that properly controlled temperature, humidity, and ventilation is often critical in maintaining the efficiency of workers. It is generally recog-

nized that electronic equipment requires proper control of the above environmental conditions, but too often, similar consideration is not given to control of the environment for the sake of workers.

VISUAL ENVIRONMENT

Perhaps the most critical environmental consideration is the visual one. Minimization of glare and reflections is especially important to insure proper seeing conditions for electronic display viewing. Solutions are not easily obtained in many cases because VDTs are used in such a wide variety of architectural situations already established and in operation. Designers must recognize potential visual environmental problems in advance and attempt to design individual equipment items to compensate for the expected problems of too

much light, glare from windows, reflections from built-in luminaires, and so forth. Filters, hoods, adjustable display angles, and display contrast controls are but a few of the techniques commonly used to combat the aforementioned problems.

LIGHTING SYSTEM DESIGN

1. Purpose
Lighting is typically provided for two main purposes: (*a*) to enable people to see what they need to see in order to perform various tasks and (*b*) to create decorative effects. It is important to differentiate between these purposes because the objectives of each are different. In some cases decorative lighting schemes may interfere with task illumination unless there is a constant awareness that the ob-

server, to see well, must have enough of the proper type of light, placed where it does the most good, and that there should be no extraneous light or reflections that might interfere with the observer's seeing process.

Seeing tasks are not always the same; thus the amount and type of lighting and the method for providing illumination should be tailored to fit the particular seeing task. A few examples will serve to illustrate how different tasks and lighting system requirements can be:

a. Reading a book or newspaper
b. Reading instruments or cathode ray tube displays in a darkened room or cockpit
c. Performing surgery or sewing on dark material
d. Inspecting for flaws in material or metal surfaces
e. Matching colors for material selection
f. Developing film in a darkroom

2. Types of Lighting
a. Natural light: The proper manipulation and use of natural light is very desirable for many applications, especially because it tends to make things we look at seem more "natural."
b. Artificial light: Artificial light is used when there is not enough natural light available for the desired seeing task and/or when it may be necessary to control the color of the light for a special seeing task.

3. Lighting Methods
The two basic types of illumination either for illuminating seeing tasks or for creating decorative lighting effects are (a) general illumination and (b) supplementary illumination

a. General illumination: General illumination should be used when the seeing requirements are broad and varied, including seeing to move about safely, observing other people or general features within the surrounding area, performing housekeeping chores, moving materials or equipment, and performing general maintenance tasks.
b. Supplementary illumination: Supplementary illumination is required when general illumination is not available and/or when the nature of the task is such that the general illumination is insufficient, is not appropriately directed, or is of such a quality or color as to make specific seeing tasks difficult to perform. Supplementary illumination also is generally used for decorative effects since it can be pre-

cisely controlled in terms of color, direction, and amount.

4. Lighting Techniques
a. Floodlighting: As the term implies, this type of lighting is usually external to the object or objects that are to be illuminated, literally flooding them with light. Floodlighting is generally used to illuminate large areas, either indoors or outdoors, but it can also be used in more concentrated, localized situations such as for illuminating a control panel, a doorway or platform, a theatre stage, a parking lot, or an airport apron. Although some control can be exercised over the flood pattern, it should be noted that the illuminated area covered by a floodlighting system will vary from the beam center to its periphery. Care must also be taken in using the floodlighting technique to consider the problems that might be associated with shadows, which are a by-product of the floodlighting technique.
b. Internal lighting: A wide variety of internally generated lighting techniques are available to make objects or visual displays "self-luminous." Typical applications for which internal lighting is appropriate include signs, panel labels, vehicle or operator console instruments and displays, shadow boards, and marker and signal lights. Internal lighting can be created in many ways, including the use of luminous phosphors, back-lighted screens, edge-lighted plastics, and end-lighted plastic "light pipes." The primary advantage of internal lighting is that more precise control can be obtained over where the light will appear.

5. Light Level
Avoid the temptation to believe that merely providing higher and higher illumination levels will result in improved seeing conditions. Although the level or intensity of light is obviously important, light can be too bright as well as too dim for the seeing task. Also remember that as light levels increase, glare problems also increase. There is usually a point below which seeing is difficult and another point above which seeing either does not improve or becomes more difficult for each type of seeing task.

6. Color
The color of the light used (i.e., for general illumination, supplementary illumination, or

signal lighting) is important from several standpoints:

a. A broad-spectrum light (white light) is desirable to make the colors of objects or people seem natural. Natural light is used as the base-line reference since it is by this reference that we observe most things during everyday living. Several types of artificial light have been created to simulate natural light, but some are better than others for certain purposes. Care must be taken to select the artificial light color that is most appropriate for the task or seeing conditions being created. Incandescent luminaires typically provide more reds and yellows, which impart a warm appearance to viewed objects. Fluorescent lights, on the other hand (depending on the specific luminaire), tend to accentuate the blue. Often, a colored material selected under fluorescent light will not look the same when the material is viewed under natural sunlight.
b. Monochromatic light (i.e., red, yellow, blue, and green) may be desirable for certain types of seeing tasks and/or for decorative purposes. Such colors are not, however, desirable for general seeing conditions; they make surface colors appear abnormal and/or impossible to recognize. Monochromatic light is particularly useful for creating color-coded displays, e.g., traffic signal lights, airport runway and taxiway lights, and warning displays.

7. Light Fixture Type and Usage
An extensive array of lighting fixtures, lamps, and materials is available for creating almost any type of lighting that is required. Some precautions need to be taken, however, in selecting commercially available fixtures. Some provide effective task illumination, but others are merely attractive to look at. When looking for appropriate fixtures off the shelf, designers should analyze the seeing task fully so that they can evaluate the merits (or, more often, the demerits) of the fixtures they are contemplating for a particular application.

An important factor to consider is the location of lighting fixtures, for even a good fixture's value can be lessened if it is improperly positioned relative to the seeing task. In addition, it is important to consider the light control requirements in terms of the possible need for dimming and/or selectively illuminating only certain fixtures at certain times during an operation.

General Illumination Guidelines

Task Requirements	Light Level, fc	Type of Illumination
Small detail; low contrast; prolonged viewing; fast, error-free response	100	Supplementary lighting fixture located near visual task
Small detail, fair contrast, close but short-duration work, speed not essential	50–100	Supplementary lighting and/or well-distributed and diffused general lighting
Typical desk and office work	40–60	General lighting with diffusing fixture directly overhead
Sports (e.g., tennis and basketball) or indoor recreational games (e.g., ping pong and billiards)	30–50	General lighting with sufficient number of fixtures to provide even court or table illumination
Recreational reading and letter writing	25–45	Supplementary lighting, positioned over reading so that page glare does not occur
General housekeeping, detail not required	10–25	General lighting
Visibility for moving about, avoiding people and furniture, and negotiating standard stairs	5–10	General and/or supplementary lighting (with care taken not to allow supplementary sources to project in the user's eyes)

Note: The above guidelines are only approximations. They are higher than some recommendations, not for seeing, but because these levels provide an additional psychological benefit as well. Levels relate to light levels measured at the primary seeing point, e.g., the desk or table surface on the floor or stair tread level.

Brightness ratios between the seeing task and the immediate surroundings should not exceed 5:1; between the task and the remote surroundings, 20:1; and between the immediate work area and any other remaining visual environment, 80:1.

Natural or white artificial light should be used regardless of the type of illumination; i.e., these levels do not apply to monochromatic light sources.

Special Situations Where Visual Dark Adaptation Is Required

Work Situations	Light Level, fc	Type of Illumination
Darkened control rooms and cockpits at night	0.01–0.4	Illumination on work surfaces
	0.03–0.10	Black and red instrument face detail (scale marks, numerals, and labels)
	0.10–1.00	Indicator lights (not to exceed 1 in, 2.5 cm square)
Ready rooms where crew members are required to maintain some dark adaptation prior to a mission at night	0.50–2.00	Illumination at tabletop
Movie theatre during film running	0.50–2.00	Illumination in aisle and immediate vestibule

Note: Red illumination is required in military workplaces and ready rooms. Low-level white light is satisfactory in theatres.

ILLUMINATION LEVEL GUIDELINES FOR MAINTAINING PROPER BRIGHTNESS CONTRAST BETWEEN VISUAL TASK AND BACKGROUND

Brightness Ratios

Comparison	Environmental Classification* A	B	C
Between tasks and adjacent darker surroundings	3:1	3:1	5:1
Between tasks and adjacent lighter surroundings	1:3	1:3	1:5
Between tasks and more remote darker surfaces	10:1	20:1	†
Between tasks and more remote lighter surfaces	1:10	1:20	†
Between luminaires and adjacent surfaces	20:1	†	†
Between the immediate work area and the rest of the environment	40:1	†	†

Note: Direct glare arises from a light source within the visual work field. It should be controlled by:

1. Avoiding bright light sources within 60° of the center of the visual field. Since most visual work is at or below the eye's horizontal position, placing luminaires high above the work area minimizes direct glare.
2. Using indirect lighting.
3. Using more relatively dim light sources, rather than a few very bright ones.
4. Using polarized light, shields, hoods, or visors to block the glare in confined areas.

*A—Interior areas where reflectances of entire space can be controlled for optimum visual conditions. B—Areas where reflectances of immediate work area can be controlled, but there is only limited control over remote surroundings. C—Areas (indoors and outdoors) where it is completely impractical to control reflectances and difficult to alter environmental conditions.

†Brightness-ratio control not practical.

LIGHT-SOURCE COLOR

The color of the light source should be chosen carefully to fit the application, as is indicated by the accompanying table.

Effect of Colored Light on Appearance of Object Color

Object Color	Red Light	Blue Light	Green Light	Yellow Light
White	Light pink	Pale blue	Pale green	Pale yellow
Red	Brilliant red	Dark bluish red	Yellowish red	Bright red
Light blue	Reddish blue	Bright blue	Greenish blue	Light reddish blue
Dark blue	Dark reddish purple	Brilliant blue	Dark greenish blue	Light reddish purple
Green	Olive green	Green-blue	Brilliant green	Yellow-green
Yellow	Red-orange	Light reddish brown	Light greenish yellow	Brilliant light orange
Brown	Brownish red	Bluish brown	Dark olive brown	Brownish orange

APPEARANCE RATINGS OF TYPICAL SURFACE COLORS WHEN VIEWED UNDER VARIOUS ARTIFICIAL LIGHT SOURCES

Color	Daylight	Fluorescent lamps				Incandescent Lamps
		Standard Cool White	Deluxe Cool White	Standard Warm White	Deluxe Warm White	
Maroon	Dull	Dull	Dull	Dull	Fair	Good
Red	Fair	Dull	Dull	Fair	Good	Good
Pink	Fair	Fair	Fair	Fair	Good	Good
Rust	Dull	Fair	Fair	Fair	Fair	Good
Orange	Dull	Dull	Fair	Fair	Fair	Good
Brown	Dull	Fair	Good	Good	Fair	Good
Tan	Dull	Fair	Good	Good	Fair	Good
Golden yellow	Dull	Fair	Fair	Good	Fair	Good
Yellow	Dull	Fair	Good	Good	Dull	Fair
Olive	Good	Fair	Fair	Fair	Brown	Brown
Chartreuse	Good	Good	Good	Good	Yellowed	Yellowed
Dark green	Good	Good	Good	Fair	Dull	Dull
Light green	Good	Good	Good	Fair	Dull	Dull
Peacock blue	Good	Good	Dull	Dull	Dull	Dull
Turquoise	Good	Fair	Dull	Dull	Dull	Dull
Royal blue	Good	Fair	Dull	Dull	Dull	Dull
Light blue	Good	Fair	Dull	Dull	Dull	Dull
Purple	Good	Fair	Dull	Dull	Good	Dull
Lavender	Good	Good	Dull	Dull	Good	Dull
Magenta	Good	Good	Fair	Dull	Good	Dull
Gray	Good	Good	Fair	Soft	Soft	Dull

Note: Good—color appears most nearly as it would under an ideal white-light source, such as north skylight. Fair—color appears about as it would under an ideal white-light source, but is less vivid. Dull—color appears less vivid. Brown—color appears to be brown because of small amount of blue light emitted by lamp. Yellowed —color appears yellowed because of small amount of blue light emitted by lamp. Soft—surface takes on a pinkish cast because of red light emitted by lamp.

GLARE AVOIDANCE AND SHADOW PREVENTION

Because everyone has at one time or another experienced seeing problems due to shadows and/or glare, it would seem unnecessary to advise designers to exercise care in designing so that these conditions will not occur. However, shadows and glare are the most commonly encountered problems in lighting system designs. The three most serious conditions are illustrated by the accompanying sketches:

Shadows are cast by the observer because of improper luminaire positioning. Obviously, the light source should not be behind the observer. Typically, the luminaire should be located slightly to the right or left so that the light comes from over the observer's shoulder. General practice suggests that the light come from over the left shoulder because most people are right-handed. A lighting mockup is recommended for determining the best location.

The light source should not be visible to the observer; i.e., the light bulb or lamp should be screened either by some type of baffle or by a distributing screen. Once again, since layout errors have a way of slipping by, use a lighting mockup to make sure that the observer cannot see the light source directly from his or her various working positions.

Ambient glare from frontal or peripheral windows and/or similarly large and bright artificial sources not only makes seeing difficult but also produces general worker fatigue. Many methods are available for dealing with this problem, including reorienting the worker, putting shades over the windows, and adding light-filtering materials to the glazing.

Although this is not always the prerogative of the lighting system designer, try to eliminate glossy surfaces in front of the observer.

Diffusing light sources (fluorescent) are recommended to minimize glare and shadow effects.

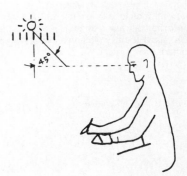

SUPPLEMENTARY LIGHTING FOR THE INDIVIDUAL WORKPLACE

Floodlighting techniques for individual workplace illumination should be designed so that the operator can select or adjust the best position for his or her seeing task; sometimes this changes.

The key consideration is to place the luminaire so that it does not cause light rays to reflect from the work surface into the operator's eyes and so that the light source (lamp) is not directly visible to the operator.

In general, the best lamp position is behind and to the side of the operator. If the operator is right-handed, the lamp should be to his or her left; if the operator is left-handed, it should be to his or her right. Thus it is necessary to provide a flexible positioning system.

The accompanying sketch illustrates a typical flexible-arm luminaire that provides a wide range of positioning possibilities.

For some special tasks, such as engraving, the light source must be placed so that the light rays actually cast shadows because these provide important guidance cues for the worker.

A low position for the light source, as shown in the accompanying sketch, is typical of the special situations in which shadow casting is advantageous.

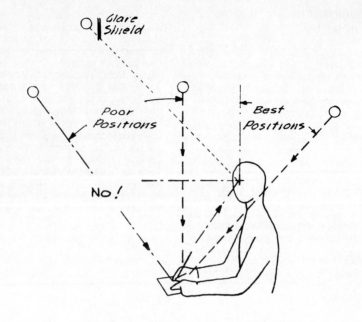

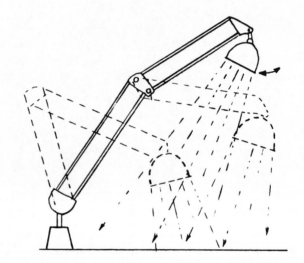

**Console Panel and Instrument
Illumination**

Illumination Guidelines for Special Military Applications

Task and Area	Illumination Level, fL	Lighting Equipment Features
Dark adaptation required	0.02–0.1	Rheostat control throughout the range; red light*
Dim-out and CRT target detection	0.02–1.0	Rheostat control throughout the range; red or white light*
Dim-out, CRTs, status boards, map boards, and console operation	1.0–3.0	Separate rheostat control for each piece of equipment; white light*
Hallways, stairways, rest rooms, and storage areas	10.0 minimum	General fixed lighting
Equipment maintenance, tape reel removal, and intermittent record keeping	25.0 minimum	General and/or supplementary fixed lighting
Classroom, drafting, teletype, and telemetry readout	35.0 minimum	General and/or supplementary fixed lighting
Fine detail, optical, machining, and assembly work	50.0 minimum	Special supplementary lighting
Critical visual tasks, e.g., surgery, small instrument repair, prolonged map work, and laboratory work using a microscope	100.0 minimum	General plus supplementary lighting

*There should be at least two lamps per instrument and/or panel label for all vehicle applications so that failure of a lamp will not leave the display completely unilluminated.

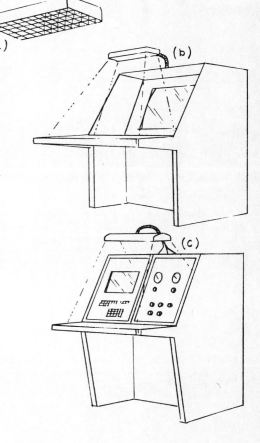

(a)

(b)

(c)

Architectural Systems

GENERAL

This section deals with human factors considerations that should be addressed during the initial conceptualization of any architectural system.

User-Oriented Conceptual Planning

Start with the user. Recognize his or her characteristics and constraints. Determine the user's needs. Create a place for the user to perform whatever tasks he or she expects to do.

Step 1: Define and examine the needs of the total user population; i.e., do not concentrate only on the primary resident, for example, but look at the needs of his or her visitors or clients and the people who will serve the primary resident in the proposed facility.

Step 2: Examine and define the various tasks that each of the above users has to perform. Determine what these tasks imply in terms of space, environmental control, supporting furnishings, and utilities.

Step 3: Explore the interactive as well as the isolative needs of the various users and their furnishings and equipment. Examine alternative arrangements to determine the most convenient organization of people, furnishings, spaces, buildings, etc.

Step 4: Create an enclosure for the most effective alternative defined in Step 3 and add appropriate partitioning to provide desired environmental control, privacy, and security.

Step 5: Select an appropriate site that will accommodate the building or buildings defined in Step 4 and locate, position, and arrange the building or buildings with respect to appropriate site and building access.

Now you are ready to examine the concept in terms of aesthetic features, including architectural style, special material effects, and landscaping. These features are generally cosmetic, and although they are important in terms of making the system pleasant to look at, they can for the most part wait until the five steps outlined above have been completed.

User Population Characteristics

Characteristic	Architectural Implication
Cultural factors	Considerable variation exists among people with respect to their cultural background, including social mores, religious attitudes, intellectual development, skill development, attitudes toward others, and where and how they live in terms of spatial features and modern technological amenities. Language differences create an important barrier to communications in many system operational settings.
Body size	People of different nationalities, as well as individuals of the same nationality, vary considerably in terms of size. There are also differences in size between children and adults, between men and women, and between members of special user populations. Differences in size impact on architectural space, including clearances and reach distances.
Mobility	The agility of various individuals varies considerably (e.g., between the young and the old and between handicapped and nonhandicapped persons), and mobility may be restricted by the garments people wear. The impact of restricted mobility on human-architectural interfaces may be critical to the operational utility of a system concept.
Strength	Very young and very old people have considerably less strength than those in the middle range, women are generally weaker than men, and handicapped persons may have virtually no strength. Architectural features that require lifting, pushing, pulling, or twisting must be tailored to the weakest member of the expected user population.
Sensory factors	Principal sensory factors associated with architectural systems relate to vision, hearing, and touch. Although only persons with so-called normal capacities may make up the expected user populations of special systems (because of operator selection restrictions), most general system concepts require consideration of the more limited capacities of elderly and handicapped individuals, especially the visually and aurally handicapped.
Motor skills	A limited number of people have superior motor skill capabilities as a result of either innate capability or training. Others are limited both innately and by lack of training. Still others are even more limited by physical handicaps.
Cognitive skills	Variation in cognitive skill occurs because of age differences, differences in education and/or technological opportunity, and innate mental handicaps. Understanding the operational aspects of the proposed architectural concept is critical to its effective use.

Defining the Needs, Characteristics, and Limitations of the Expected User Population

A clear definition of the needs, characteristics, and limitations of the expected user population should be sought before any architectural concept is frozen. Otherwise, the system may not be completely compatible with user expectations or abilities to use the system effectively. Because people are different, it is a mistake to assume that a system can be designed for the so-called average person. Understanding these differences and accommodating the proposed concept to them are vital to the eventual operation of the system. The accompanying table lists the important characteristics that should be addressed in defining the user population.

User Efficiency

It is often said that "user efficiency does not sell products—appearance does." From a human factors point of view, however, efficiency is of prime importance to the eventual effectiveness of any system. The table provides suggestions regarding user efficiency variables that should be considered carefully during the conceptual phase of any architectural system development.

User Efficiency

Parameter	Variables
Vision	What a person sees clearly establishes the basic input to that person. His or her use response depends on how well the architectural concept implies what the designer intends the user to do with it. The critical variables include the following: 1. Visibility: Are critical features in sight, or are they obscured by intervening elements, glare, or shadows? 2. Legibility: Are critical features clear, or are they distorted by lack of contrast, parallax, exaggerated embellishment, or illusory geometrics? 3. Conspicuousness: Are features that are important to detecting, recognizing, and understanding lost in the background? 4. Recognizability: Are features natural, familiar, and/or similar to the observer's expectations, or are they distorted or purposely made to look like what they are not?
Hearing	What people hear not only affects their ability to communicate but may also affect their general capacity to perform other tasks. The critical variables include the following: 1. Audibility: If certain sounds must be heard, the acoustic environment must be designed to carry the sounds and not block them. 2. Intelligibility: The acoustic environment must be designed so that it will not distort the sounds intended for the listener. 3. Signal-to-noise ratio: The combined communications and acoustic system must be designed to maximize the probability that extraneous noises will not obscure the desired sound signal. 4. Noise annoyance: Adequate noise attenuation must be provided to minimize the possible deleterious effects that an annoying noise can have on individual task performance.
Stability	How well a person performs ambulation or biomechanical or other manipulative tasks depends on the stability-aiding elements of the architectural system and/or the possible impediments designed into the system. In addition, there are critical visual interactions that may add to the instability of the user. Among the typical features to examine are the slope of floors, walkways, stair treads, handrails, and door thresholds. Structural vibration also impacts on user stability.
Mobility	How well people perform dynamic tasks in which they must move their bodies and limbs depends both on the clearances provided around their task envelope and on the supporting area provided to maintain stability.
Convenience	How well people perform various tasks depends to a great extent on how conveniently they can move from one place to another. This requires careful consideration of functional relationships, the sequence of events, time constraints, and emergency demands in order to create a logical and energy-saving arrangement of spaces and activities within spaces. Lack of convenience not only reduces immediate user efficiencies but also may add to fatigue and possible operator failures.

Handicapped Users

Special consideration should be given to the needs of the handicapped when it is obvious that they too can be expected to utilize a proposed architectural system. The following description of typical handicaps provides a general idea of how the architectural system will have to be modified so that the handicapped can use it effectively.

1. The blind and partially blind: Such individuals must depend on sound and/or touch in order to locate, identify, and interpret their physical surroundings. Those who have been blind from birth must depend on touch since they have never seen alphanumeric or other symbols. The partially blind require that visual patterns and brightnesses be maximized.

2. The deaf and partially deaf: Such individuals have less difficulty within the architectural context because they receive most of their information visually. However, they cannot hear warning sounds, which may be the only signal of impending danger.

3. The orthopedically handicapped: Most current standards for the handicapped emphasize special designs related to the use of wheelchairs because of the constraints that such devices place on the mobility of their users. However, there are many other considerations that too often are neglected by the designer. Among these is the fact that many orthopedically handicapped individuals have hand deformities that prevent them from grasping, from applying force, from reaching, etc. In addition, consideration should be given to the person who may be required to use one or more crutches—an entirely different set of constraints.

Special note: Many aged persons have many of the above handicaps, which highlights the importance of not accepting the "average person" approach to architectural design. Also, it is highly probable that many facilities may have to be used by individuals who are only temporarily handicapped. Many of the special features required by the handicapped may also be helpful to the average user, e.g., more maneuvering room, better handrails, and better handles. On the other hand, care must be taken to avoid creating aids for the handicapped that may cause difficulties for nonhandicapped users.

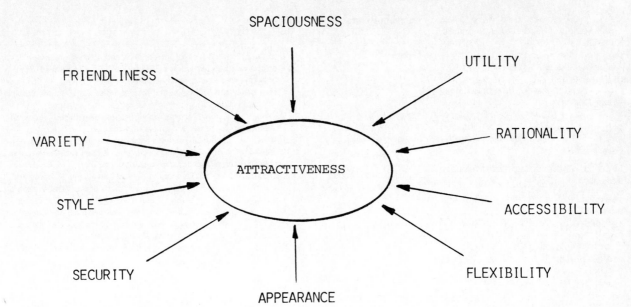

Psychological Considerations

The architect is usually concerned about whether the user will be attracted by the design of the community, home, building, or other structural edifice—not only when it is first observed, but also as it is occupied over a longer period of time. The model shown above may be useful in considering the more important ingredients that play a part in determining user reaction.

The accompanying table expands on these ingredients, listing common responses given by typical research test subjects who were asked to verbalize their reactions to architectural environmental features.

In general, one should seek a more or less middle ground in trying to create a balanced combination of the above factors. However, the adjustment between descriptors will be different for different types of architectural systems. For example, the objectives for a satisfactory home environment are not necessarily the same as those for a satisfactory office or factory environment. By the same token, similar adjustment is required for subsystems within the home, office, or factory; e.g., the psychological needs in the bedroom are different from those in the bathroom, and the needs of the production department are different from those of the company library.

Personal Space

Individuals perceive their relationships with others in terms of the distance between themselves and the people they can see. At least four distinct territorial categories have been defined by various researchers:

1. Public: Those areas where the individual has freedom of access, but not necessarily of action.
2. Home: Those areas where the regular participants have regular freedom of behavior and a sense of control over the area.

Semantic Descriptors for Assessing Observable Physical Features in Architecture

Factor Categories	Descriptor Scaling Examples	
Spaciousness	Generous	Cramped
	Ample	Limited
	Empty	Crowded
Friendliness	Warm	Cold
	Intimate	Detached
	Relaxed	Stiff
Variety	Stimulating	Boring
	Dynamic	Static
	Diverse	Monotonous
Utility	Purposeful	Unnecessary
	Efficient	Confusing
	Practical	Frivolous
Rationality	Organized	Uncoordinated
	Logical	Confusing
	Simple	Complex
Flexibility	Adjustable	Constrained
	Mobile	Fixed
	Expandable	Contained
Accessibility	Open	Closed
	Direct	Indirect
	Formal	Casual
Security	Familiar	Unknown
	Safe	Uncertain
	Protected	Exposed
Appearance	Graceful	Awkward
	Contemporary	Obsolete
	Meaningful	Obscure

3. Interactional: Those areas where social gatherings may occur. An invisible boundary and a territorial claim are implicit, though not officially promulgated by the people present.
4. Body: The area immediately surrounding the individual's body. This area is most private and inviolate to the individual.

Although absolute distance criteria are probably not pertinent, the following approximations are useful for architectural space consideration:

1. Intimate—0 to 18 in (46 cm): People desire involvement with each other, both physically and emotionally.

2. Casual-personal—30 to 48 in (76 to 122 cm): People will accept minimal contact (a handshake), but wish to retain freedom from domination.
3. Social-consultative—7 to 12 ft (2.1 to 3.7 m): People desire to maintain a certain formality and the freedom to break away when desired.
4. Public—30 to 1500 ft (9 to 450 m): People want to be able to remain unnoticed and to flee or be independent from the crowd.

Many factors are related to the individual's need for personal space:

1. The desire to converse privately in a subdued voice
2. The desire to interact intimately with a loved one
3. The desire to avoid physical contact with another person or the offensive odor of another person
4. The desire to see the eyes of another person clearly
5. The desire to view another person completely at a single glance
6. The desire to be an observer, but not an active participant

As these factors imply, architectural features can bear importantly on user reaction to space.

Perceptual Quality of the Designed Environment

The following principles, although not easily quantified, provide a further, subjective scaling of design characteristics that may reduce the probability of negative response to the perceived environment:

1. Order: Most people are impelled to seek order and understanding, but they also need sufficient variety to be stimulated by what they see.
2. Outline: The outline of the "whole" should represent grace and balance, not awkward angularity, overpowering massiveness, or unintentional asymmetry.
3. Identifiable references: Environmental references—i.e., paths, edges, districts, nodes, landmarks, runs, mergings, portals, areas, volumes, and acoustic divisions—should be clear.
4. Functional form: Space should appear "positive" rather than "negative"; i.e., it should seem to have been purposefully designed, rather than left to chance.
5. The whole versus a sequential experience: Perceptual confidence comes from an understanding of the whole, as opposed to a sequential experience, which leads to continuing assumptions and doubts.
6. Familiarity: An impression of security based on the repetition of familiar patterns should be created, but without incurring boredom or monotony.
7. Reliability: Visual illusions that could lead

Human-Architectural Interfaces

Interface	Consideration
Site location	Distance to related facilities: residential, commercial, and industrial areas; airport and bus terminals; etc. Accessibility (or nonaccessibility) to highway, roadway, street, rail, and waterway systems: vehicular or pedestrian, etc. Environmental factors: noise, air pollution, possible natural disasters, traffic problems, and aesthetic factors
Site amenities	Parking, public transit access, walk-in customer exposure, landscape, illumination, security, and emergency access
Building or buildings, external	Number, size, and location; entrances (number, location, and type); identification and illumination; special amenities (utilities, sidewalks, and stairs); external maintenance requirements; etc.
Building or buildings, internal	Compartmentalization (number, size, and arrangement); doors and windows (type, number, size, and location); stairs, ramps, elevators, and escalators; heating, air conditioning, ventilation, and illumination; built-in features (cabinets, plumbing fixtures, and electrical and pneumatic systems); floor covering; safety equipment; communications systems; acoustics; and internal maintenance requirements
Safety	Fire protection, emergency escape, first aid, and disaster control

to incorrect assumptions and loss of confidence on the part of the observer should be avoided.
8. Cultural identity: Cultural differences reflect individual needs to identify with the traditional, as opposed to challenges to keep up with what is fashionable.
9. Aesthetic objective: The aesthetic objective should be relevant to human needs rather than architectural monuments; i.e., it should provide psychosocial values with which to identify, it should express the user's individuality (not the designer's), and it should provide general perceptual enrichment.

Definition of Critical Human-Architectural Interfaces

Although the specific human-architectural interfaces and the level of criticality of each interface may vary from one system to another, those listed in the table above should be considered during the conceptualization of any new system.

Considering Ease of Maintenance during the Planning Stage of Design

If it is considered at all during the planning stage, facility maintenance traditionally becomes a question of: How can this facility be maintained as we have designed it? In other words, too often the question of ease of maintenance does not come up when key architectural configuration decisions are being made. In spite of this traditional attitude, all facilities have to be maintained, and by human beings. The following key maintenance functions should be part of any design concept trade-off analysis:

1. Daily housekeeping: cleaning floors, walkways, windows, walls, ceilings, etc.

2. Periodic inspection and repair: inspecting and repairing windows, roofs, walls and woodwork, hot water heaters, plumbing, etc.
3. Periodic refurbishment: repainting exterior and interior surfaces, replacing roofs, replacing plumbing fixtures, etc.
4. Landscape maintenance: watering lawns and shrubs, removing trash, etc.

Common Human Factors Problems Associated with Maintenance-Related Design

1. One cannot get to the spot that requires inspection, adjustment, cleaning, removal, replacement, or refurbishment.
2. There is insufficient space to do the job once a person has reached or located the maintenance problem.
3. There is insufficient illumination to see what needs to be seen.
4. There is a lack of appropriate service connections to enable use of the necessary tools at the work site.
5. The device to be repaired or replaced is buried into the structure, requiring major destruction and eventual repair.
6. Main service shutoffs (e.g., water, electricity, or gas) are variously distributed, hidden, and/or inaccessible, requiring an inordinate amount of time to find them.
7. The composite land-site–structural relationship precludes the normal and safe use of common maintenance aids such as ladders or scaffolds.

Architectural Safety

An objective during conceptual planning should be to create an environment in which the user can be as safe as possible. Although this is a tall order, many of the accidents that frequently occur in homes, offices, schools, factories, and elsewhere are due as much to the facility design as they are to user errors.

Communications and Information Processing

GENERAL

Communications Systems

In conceptualizing a communications system, whether it is large and complex (such as a worldwide satellite system) or relatively simple and limited (such as an internal closed-circuit television system in a factory), the common denominator should always be the human need. That is, communications between people should be analyzed and planned on the basis of human task needs and thus should be designed to meet these needs in the most effective manner. This requires special consideration of human input-output characteristics, or, if you will, human channel capacities and sensory channel sensitivity. In developing a communications system concept, consider at least the following:

1. Communications mode (i.e., visual, auditory, tactile, and/or some combination)
2. Communicating environmental constraints (i.e., noise, environmental interference, and the need for security or privacy)
3. The time factor (i.e., how much informa- tion must be transmitted within some prescribed time frame)
4. Reliability requirements (i.e., whether you can afford to repeat or lose some of the information without degrading system performance objectives)
5. Convenience (i.e., whether immediate accessibility is critical)
6. Multiuser requirements (i.e., whether the same communication is important to several users simultaneously)
7. Personal innuendo (i.e., the value of face-to-face expression)
8. International language (i.e., whether it is critical that people with differing language backgrounds understand the communication)
9. Freedom from interruption (i.e., whether it is critical that a given communication not be interrupted prior to its complete transmission)
10. System cost, both the initial cost and the cost of maintenance
11. The effectiveness of the human-machine interface design (i.e., ease of use of handsets, microphones, earphones, controls, etc), considering the environmental and other constraints involved in use of the interface (temperature, noise, vibration, clothing restrictions, etc.)

General Considerations for Development of Communications System Concepts

The following discussion is limited primarily to visual and auditory communications between people within the working community, as opposed to communications for the purpose of recreation or entertainment. A primary objective should be to create a system that supports other human task activities, e.g., the exchange of information among individuals and the transmission of instructions, commands, warnings, orders, etc. Such systems should be based on the capabilities and limitations of the human as well as on the information transfer needs of the users. Consider the following:

1. Type of information (spoken messages, pictures, coded messages, digital data, etc.)
2. The time factor (i.e., how rapidly the information must be transmitted)
3. Security (i.e., whether the information is to be accessible to only one person or to anyone who might be within hearing or seeing distance of the receiving device)
4. Number of intended recipients (i.e., whether the message is for general broadcast for the benefit of a large group of recipients)

141

5. The receiving environment (i.e., whether certain interferences at the receiving station may mask the incoming communication)
6. Reliability requirements (i.e., whether the situation permits repetition of the message if it is not understood the first time)
7. Quality (i.e., whether the information requires high fidelity in order to be fully understood)
8. Language (i.e., whether it is critical that people of differing language backgrounds understand the communication)
9. Personal factors (i.e., how much value there is to seeing facial expressions or hearing voice inflections)
10. Directionality (i.e., how important it is to associate the transmission with a direction or source)
11. Message capture (i.e., whether it is necessary to have a record of the transmission)
12. Condition of the addressee (i.e., whether he or she is busy, is using his or her hands or eyes, or is under special physical stress, such as high *g* forces)

Communications hardware technology has reached the state at which the planner can prescribe almost any level of fidelity or reliability and almost any mode. Therefore, the choice is limited primarily by cost. However, since the primary purpose is to allow people to communicate with one another, systems concepts should generate from human needs and from basic human sensorimotor characteristics and limitations. In defining the characteristics of any proposed system, consider the following:

For Auditory Communications

1. Loudness (intensity)
2. Pitch (frequency)
3. Timbre (signal and voice profile)
4. Signal-to-noise ratio
5. Signal pattern (duration, intermittency, and frequency shift)
6. Sound location
7. Public as opposed to private

For Visual Communications

1. Brightness
2. Color
3. Contrast
4. Duration
5. Pattern and shape
6. Alphanumeric legibility

The following are some of the important human factors objectives for communications systems:

1. Detectability: The intended receiver can sense the signal.
2. Recognizability: The intended receiver can tell what the signal is.
3. Intelligibility: The intended receiver can tell what the signal means.

4. Conspicuousness: The signal is attention-getting.

Although communications systems designers obviously are concerned with the basic transmission medium (i.e., they must select radio, radar, sonar, or another basic scheme for the system), the human factors relate primarily to the immediate interface between the system and the sender, listener, or visual observer. Although tactile communication interfaces may have some application in isolated instances, the primary questions are whether the message originator should transmit by voice, encoded aural signal, or written or graphic means; whether the message will be received by aural or visual means; and whether there is increased value (e.g., reliability) if the message is transmitted using both visual and auditory modes.

1. When to Use Auditory Communication
a. When the message is simple
b. When the message is short
c. When the message does not necessarily have to be referred to later
d. When the message deals with events in time
e. When the message calls for immediate action
f. When the visual system is already overloaded
g. When the intended receiver cannot look at a visual display or may have to move about continually
h. When the receiving environment is incompatible with seeing, (e.g., when it is too bright or when it is necessary to avoid visual signals that could interfere with the intended receiver's adaptation to the dark)
i. When the intended receiver's visual capacity is impaired as a result of a dynamic stress (e.g., high *g* forces)
j. When the intended receiver's eyes may be closed (e.g., when he or she is asleep)
k. When a visual signal strength is borderline and the addition of a coincident auditory signal would provide redundant, reinforcing confirmation

2. When to Use Visual Communication
a. When the message is very complex
b. When the message is long
c. When the message needs to be referred to later
d. When the information deals with location in space
e. When the message does not necessarily call for immediate action (e.g., when the receiver needs to finish some other task)
f. When the auditory system of the receiver is overburdened
g. When the receiving environment is too noisy to ensure reliable receipt of an aural message
h. When the receiver will remain in a position where he or she can continue to watch the visual display
i. When the information characteristics cannot be reliably described by words
j. When face-to-face communication may provide added understanding between communicators

3. When to Use a Private Rather than a Public Interface
An important decision in developing any communication system concerns whether, and under what circumstances, to use individual rather than group visual or auditory displays. Although there are no hard-and-fast rules, the following should be considered:

a. Use group display interfaces when:
   (1) Several people should receive the same information simultaneously.
   (2) Several people should all receive the information in the same format.
   (3) Some people require general information at the same time that they are busy sending or receiving individual communications.
   (4) The transmitter may not know who the persons are who should receive the information.
   (5) The individual channel that the transmitter wishes to use is temporarily unavailable.
   (6) The receiving individual is temporarily away from his or her individual communication interface.
   (7) Input to a general status display is being generated by several individuals on a continuing basis.
   (8) It is inconvenient for the required number of listeners or viewers to gather around an individual display device.
b. Use individual display interfaces when:
   (1) The transmission involves only two persons.
   (2) The information need not be shared with any other persons.
   (3) The transmission is or should be private.
   (4) The information is being modified by the individual for his or her own benefit.
   (5) The individual needs to record the information for future reference.

4. When to Use Handsets, Intercoms, or Public Address Systems
Although there are obvious overlaps in the manner in which handsets, intercoms, and public address systems can or should be used, the following are guidelines for making this decision:

a. Handsets should be used primarily when privacy is desired.
b. Intercoms should be used primarily when the individual is too busy to look up a number and dial it.
c. Public address systems should be used primarily for making public announcements.

ALARMS

**Guidelines for Selecting Auditory
Alarm Systems***

Types of Alarms, Their Characteristics and Special Features

Alarm	Intensity	Frequency	Attention-getting ability	Noise-penetration ability	Special features
Diaphone (foghorn)	Very high	Very low	Good	Poor in low-frequency noise Good in high-frequency noise	
Horn	High	Low to high	Good	Good	Can be designed to beam sound directionally Can be rotated to get wide coverage
Whistle	High	Low to high	Good if intermittent	Good if frequency is properly chosen	Can be made directional by reflectors
Siren	High	Low to high	Very good if pitch rises and falls	Very good with rising and falling frequency	Can be coupled to horn for directional transmission
Bell	Medium	Medium to high	Good	Good in low-frequency noise	Can be provided with manual shutoff to insure alarm until action is taken
Buzzer	Low to medium	Low to medium	Good	Fair if spectrum is suited to background noise	Can be provided with manual shutoff to insure alarm until action is taken
Chimes and gong	Low to medium	Low to medium	Fair	Fair if spectrum is suited to background noise	
Oscillator	Low to high	Medium to high	Good if intermittent	Good if frequency is properly chosen	Can be presented over intercom system

Summary of Design Recommendations for Auditory Alarm and Warning Devices

Conditions	Design recommendations
1. If distance to listener is great	1. Use high intensities and avoid high frequencies
2. If sound must bend around obstacles and pass through partitions	2. Use low frequencies
3. If background noise is present	3. Select alarm frequency in region where noise masking is minimal
4. To demand attention	4. Modulate signal to give intermittent "beeps" or modulate frequency to make pitch rise and fall at rate of about 1–3 cps
5. To acknowledge warning	5. Provide signal with manual shutoff so that it sounds continuously until action is taken

*C. T. Morgan, J. S. Cook, A. Chapanis, and M. W. Lund (eds.), *Human Engineering Guide to Equipment Design*, McGraw-Hill Book Company (*1st Ed*), New York, 1964.

HUMAN INFORMATION-PROCESSING CAPABILITIES

Humans can transmit only about 5 to 10 bits of information per second (b/s). They can transmit about 2 b/s when the stimuli they receive are fairly well structured, although this can often be doubled by adding appropriate coding or anchoring of the input. The relationship between bits per item and the bits-per-second limitation depends on what is referred to as "attention switch time" (e.g., 0.1 to 0.2 s). Thus an operator can accept no more than two or three items of data per second.

As the speed of information-processing demand increases, the number of errors typically increases. Thus the overall information transfer rate tends to remain constant at about 10^{-2} b/s. Through careful input display design, however, this rate can be increased to 10^{-3} or even 10^{-6} under good conditions. The following error classification scheme, which was suggested by Kidd,* helps characterize types of information-processing errors and why they occur:

1. Failure to detect a signal: input overload or underload and/or actual interference
2. Misidentification: insufficient cues
3. Improper weighting of informational factors and/or selection of input factors: poor or inadequate conceptualizations or evaluation of action choices
4. Action failure: A wrong action at the right time or a right action at the wrong time

INFORMATION STORAGE CAPACITY

Information storage is of two distinct types: long-term and short-term. Short-term memory storage capacity is generally limited to about 30 lb or eight individual items. The human generally organizes stored information in terms of sensory modality (visual, auditory, etc.). The most significant storage problem occurs because of the potential interference between old ("held") information and new items that present themselves during the holding period. This accounts for the frequent "reversal errors" in information processing. As a general principle, human memory is more effectively utilized as a means of orienting and sequencing information than as a depository for isolated data or symbolic items.

INPUT CAPACITY

The total sensory input capacity of the human system is about 10^9 b/s (as compared with output capacity of 10 b/s). The five basic input categories for the visual channel are:

Relative position
Shapes
Brightness
Color
Movement

*J. S. Kidd, "Some Sources of Load and Constraints on Operator Performance in a Simulated Radar Air Traffic Control Task," USAF WADD TR-60-612, 1961.

Skill Breakdown

Effect	Reaction
Failure of selective attention fails	Narrowing of attention
Perceptual disorganization	Reduction in the size of the data sample and actions upon it and reduction in filtering efficiency
Cumulative disruption	Temporary halts (complete stop-pages) with start-overs

For the auditory channel they are:

Pitch
Loudness
Rhythm
Timbre (the quality that allows one to distinguish different voices, instruments, or special auditory displays)

DECISION MAKING

Three basic kinds of information must be available for an operator to make good decisions:

Information regarding policies and objectives
Information regarding possible alternatives and consequences
Information about the current state of the system

It is important to recognize that, in making decisions, people pursue two different courses of action:

1. Evaluating likely outcomes
2. Keeping options open as long as possible

STRESS SUSCEPTIBILITY

Human information processing is subject to a variety of stresses that may affect the efficiency with which information is received, processed, and acted upon. Two basic factors are important to consider:

1. The state of arousal of the human system (alertness)
2. Potential skill deterioration due to disorganization, cumulative disruption, and/or fatigue

The effects of, and reactions to, skill breakdown have been categorized as shown in the table at the top of the page.

INFORMATION DISPLAY IN RELATION TO DECISION MAKING†

Although research has shown that, on the basis of information acquisition only, added display complexity generally degrades operator performance, decision-making performance is not always similarly affected:

†After W. T. Singleton, "The Ergonomics of Information Presentation," *Applied Ergonomics*, vol. 2, no. 4, pp. 213–220, 1971.

The relationship between "fact density" and decision adequacy may be curvilinear; i.e., where there is a decreasing slope under low-density conditions and/or an increasing slope under high-density conditions, actual decision-making performance may improve. For example, although complexity of information acquisition degrades overall performance when fact density is low, decision-making performance may actually increase with increases in fact density.

When information is compressed, an inverted "U-shaped" relationship occurs, with moderate levels of compression producing the best decision-making performance; i.e., the critical variable is not merely symbol count, etc., but also the subjective weighting of the particular facts being compressed.

At levels of low fact density, high compression, and high display clutter, simple coding (e.g., color) may be superior to more elaborate encoding combinations. However, at other levels of fact compression and clutter, a double coding is usually more effective (e.g., color plus size).

Subject motivation (incentive) works only at low informational levels. At high levels of motivation, degraded performance is often mediated by scattered attention.

Increases in perceptual clutter sometimes increase performance rather than degrading it, as one might expect, because we are used to working in a particularly cluttered type of information-transfer environment. The more random an irrelevancy, the more performance is facilitated because of our need to have irrelevancies for a "figure-ground" decision-making background.

It is important, however, to recognize the importance of input validity to decision making, since humans quickly recognize and react to the futility of trying to decide on information that destroys or degrades the quality of their decisions.

**AMOUNT OF INFORMATION IN
ABSOLUTE JUDGMENT FOR VARIOUS
STIMULUS DIMENSIONS**

Stimulus Dimension	Number of Levels Which Can Be Discriminated	Bits of Information Transmitted*
Audition (single dimension)		
Pure tones	5	2.3
Loudness	5	2.3
Audition (combination of six variables including frequency and intensity, on-time fraction, duration, spatial location, and rate/interruption	150	7.2
Vision (single dimension), e.g., pointer and linear scale	9	3.1
Visual size	7	2.8
Hue	9	3.1
Brightness	5	2.3
Vision (combinations):		
Size, brightness, and hue	17	4.1
Hue and saturation	15	3.9
Position of dot in square	24	4.6
Odor (single dimension)	4	2.0
Odor (combination, e.g., kind, intensity, and number)	16	4.0
Taste:		
Saltiness	4	1.9
Sweetness	3	1.7

*Lack of correspondence of bit values is due to rounding to the nearest whole number.

**VISUAL AND AUDITORY CODING
METHODS**

Summary of Certain Visual and Auditory Coding Methods

(Numbers refer to number of levels which can be discriminated on an absolute basis under optimum conditions.)

Alphanumeric	Single numerals, 10; single letters, 26; combinations, unlimited. Good; especially useful for identification; uses little space if there is good contrast. Certain items easily confused with each other.
Color	Hues, 9; hue, saturation, and brightness combinations, 15–24. Particularly good for searching and counting tasks; poorer for identification tasks; trained observers can use many codes (up to 24). Affected by some lights; problem with color-defective individuals.
Geometric shapes	15 or more. Generally useful coding system, particularly in symbolic representation; good for CRTs. Shapes used together need to be discriminable; some sets of shapes more difficult to discriminate than others.
Visual angle	24. Generally satisfactory for special purposes such as indicating direction, angle, or position on round instruments like clocks, CRTs, etc.
Size of forms (such as squares)	5. Takes considerable space. Use only when specifically appropriate; preferably use less than 5.
Visual number	6. Use only when specifically appropriate, such as to represent numbers of items. Takes considerable space; may be confused with other symbols.
Brightness of lights	3–4. Use only when specifically appropriate. Preferably limit to two levels; weaker signals may be masked.
Flash rate of lights	4. Limited applicability. Preferably limit to two levels; combination of individual flashes and controlled time intervals may have special application, such as lighthouse signals and naval communications.
Sound frequency	5. For untrained listeners, use less than 5 levels; space widely apart, but avoid multiples and low and high frequencies; intensity should be 30 dB above threshold. Frequency changes easier to detect than single frequencies; combinations usually require training except for clearly distinguishable sounds such as bells, buzzers, and sirens.
Sound intensity	4. Preferably use less than 4. Intensity changes easier to detect than single intensities; for pure tones restrict to 1000–4000 Hz, but preferably use wide band.
Sound duration	Use clear-cut differences, preferably 2 or 3.
Sound direction	Difference in intensity to two ears should be distinct; particularly useful for directional information (i.e., right versus left).

GUIDELINES RELATIVE TO INFORMATION-PROCESSING OPTIMIZATION

Multidimensional coding typically results in more information transmission than single-dimension coding.

Information-processing capacity is adversely affected by both load and speed; i.e., the number of sources should be limited.

When time sharing of sensory inputs occurs, signals should be separated temporally (preferably by 0.5 s or more).

When operators can control the input rate (are self-pacing) and have some method of identifying the more important input (if they have a choice), auditory signals are generally more durable than visual signals.

Simultaneous presentation of the same information via two sensory modalities usually increases the probability of reception.

Signals presented on one channel (such as audition) can serve as cues to facilitate the use of another channel.

Time sharing of visual tasks adversely affects primarily those tasks with greater uncertainty and those which depend on short-term memory; thus speed stress does not affect all tasks equally.

INFORMATION TRANSFER FACILITATION

Certain information transfer facilitators are at the disposal of the designer and should be utilized whenever possible to maximize the human's processing response. The accompanying table reviews some of the key facilitators that can be used.

Facilitator	Remarks
Familiar patterns	Present information to operators in a form that is already familiar to them.
Visualization	Whenever practical, utilize the natural tendency of most individuals to try to visualize, even if the presentation mode is via channels other than the visual one.
Context	Provide as complete a picture as possible in a logically organized manner.
Normal relationships	Present informational components and component-observer relationships that are natural (alphanumeric characters right side up, high-low values oriented according to specific use format, etc.).
Minimal extrapolation	Avoid requiring the observer to extrapolate; i.e., present quantities, values, and patterns in their intended form.
Reference bases	Provide a continuous reference (a scale value, a map, a comparison background, etc.).
Timing	Time inputs to avoid overload and delay; present them in a logical sequence and provide immediate feedback to the operator's responses or queries.
Noise	Control input competition, interference, and distractions.
Conspicuousness	Emphasize informational cues via adequate intensity, contrast, and special encoding to ensure maximum arousal and attention.
Expectation	Anticipate the operator's mental set and prepare the operator to receive the information.
Meaningfulness	Clarify the value, necessity, and urgency of the information transfer.

Note: For additional enlightenment with regard to human information processing, see D. E. Rumelhart, *Introduction to Human Information Processing,* John Wiley & Sons, Inc., New York, 1977.

Human Engineering Activities and Management

MANAGEMENT

Management of System Analysis Activities

The following materials provide guidelines for organizing any development project in order to perform system analyses in a timely and effective manner. The key to proper management is to plan each of the analysis activities so that they all proceed concurrently, but to include appropriate iterations to allow each analysis to influence the others.

Fundamental Cycle of the System Engineering Management Process

The accompanying diagram* illustrates a process for defining a system on a total basis so that the design will reflect requirements for equipment, computer programs, facilities, procedural data, and personnel in an integrated fashion. It provides source requirement data for the development of specifications, test plans, and procedures and the backup data required to define, contract, design, develop, produce, install, check out, and test the system.

Step 1: Identify system requirements and translate these into basic functional requirements, i.e., statements of operation. These should be portrayed in the form of top- and first-level functional flow block diagrams to portray sequential and parallel interactions of functions. It is not necessary to proclaim a solution at this time, but simply to understand its use.

Step 2: Analyze functions and associated criteria and translate these into design requirements in sufficient technical detail to provide criteria for (a) designing equipment and/or computer programs and defining fa-

*After Air Force Systems Command, *Systems Engineering Management Procedures*, AFSCM 375-5, 1966.

cility equipment and intersystem interfaces and (b) determining requirements for personnel, training, training equipment, and procedural data. These requirements should be recorded on requirements allocation sheets and time-line sheets.

Step 3: Concurrently with step 2, conduct system design engineering studies to (a) determine selection of alternative functions and function sequences; (b) determine design, personnel, training, and procedural data requirements imposed by the functions; (c) determine the best way to satisfy design requirements; and (d) select the best design approach for integrating design requirements into contractual end items (CEI) of

equipment, computer software programs, firmware, and/or operator/maintainer/user roles.

Step 4: Prepare CEIs in terms of specific performance, design, and test requirements.

Note: The above steps are iterative in the sense that, as each step proceeds, analytic information develops that suggests modifications to previous steps, which in turn require successive modifications of later steps. The key to success in this process is thorough documentation so that all members of a design team can remain apprised of all results of analysis, decisions, and current conceptual status.

FUNDAMENTAL CYCLE OF THE SYSTEM ENGINEERING MANAGEMENT PROCESS

STEP 1	STEP 2	STEP 4
SYSTEM REQUIREMENTS TRANSLATED INTO FUNCTIONAL REQUIREMENTS	FUNCTIONS ANALYZED AND TRANSLATED INTO REQUIREMENTS FOR DESIGN FACILITIES, PERSONNEL TRAINING AND PROCEDURAL DATA	REQUIREMENTS INTEGRATED INTO CONTRACT END ITEMS-AFSC'S-TRAINING COURSES - PROCEDURAL PUBLICATIONS
FUNCTIONAL FLOW BLOCK DIAGRAMS (ATCH 1)	RAS & TIME LINE SHEETS (ATCH 1)	DESIGN SHEETS (ATCH 1)

STEP 3

SYSTEM DESIGN ENGINEERING TRADE-OFF STUDIES TO DETERMINE REQUIREMENTS AND DESIGN APPROACH
TRADE STUDY REPORTS (ATCH 1)

149

System Engineering Documentation and Application Criteria*

Documentation	Description	Basic Purpose	Application
Functional flow block diagram	Identifies and sequences the system and system element functions that must be accomplished in order to achieve system project objectives.	1. Facilitates development of system requirements in view of the basic operations that must be accomplished to achieve system objectives. 2. Develops the basis for establishing intersystem functional interfaces, as well as identifying system relationships.	All system programs and projects for which the definition or acquisition phase is applicable or directed. Also selected projects in the advanced development and exploratory development elements.
Requirement allocation sheet (RAS)	Defines the requirements and constraints pertaining to each of the flow diagram functions and apportions these requirements to equipment, facilities, personnel, and procedural data.	1. Facilities development of system element requirements on a functional, system basis rather than purely a hardware basis. 2. Facilitates correlation of hardware, computer programs facility personnel, and procedural data requirements to the functions these system elements are accomplishing. 3. Identifies trade-off studies required to determine more detailed system definition.	
Trade study report	Documents the trade-offs and back-up rationale pertaining to the functional diagram and requirements developed on the RAS, design sheet, schematic, time-line sheets, and other system engineering, documentation.	Facilitates (1) systematic consideration of all possible solutions in view of defined system constraints and (2) selection of the best solution.	
Time-line sheet	Presents system functions against a time base in their required sequence of accomplishment.	1. Facilitates cognizance of time element and specific sequence parallel relationships between functions in the development of system element. 2. Used to evaluate design effectiveness in terms of reaction time, performance time, maintenance down-time, equipment and personnel utilization time.	
Schematic block diagram	Schematically identifies and represents hardware computer programs and facility subsystem and item component functional interfaces and interrelationships.	1. To facilitate development of hardware and computer programs facility items in view of constraining interfaces. 2. Used to allocate requirements developed on the RAS contract end item.	
Design sheet	Identifies hardware and computer program and facility end item performance design, test requirements. Becomes section 3 and 4 of part detail specification. (reference AFSCM 378-1)	Defines hardware computer program and facilities performance, design, and test criteria on an end item basis.	
Facility interface sheet	Identifies functional and physical interfaces between equipment and facilities on an end item basis.	Supplements the RAS to collect and further define specific equipment interfaces with the facility on an end item rather than functional basis.	Any system programs and projects for which a definition or acquisition phase is applicable or directed and that will result in equipment that has complex interfaces with facilities.
End item maintenance sheet (manual)	Summarizes maintenance requirements on a specific end item, subassembly, and component basis.	Facilitates systematic and complete development of maintenance requirements for each system, end item, and subassemblies component.	Any system programs or projects for which a definition or acquisition phase is applicable or directed and that results in relatively complex, nonstandard hardware, facilities, or computer program development which will require major logistic support.

*Air Force Systems Command, *Systems Engineering Management Procedures,* AFSCM 375-5, 1966.

(continued)

System Engineering Documentation
and Application Criteria *(continued)*

Documentation	Description	Basic Purpose	Application
Maintenance sheets (automated)	Summarize maintenance requirements on a specific end item, subassembly, and component basis. Provides data for configuration management, computer program and detail maintenance data elements. May be modified for manual use.	1. Facilitates systematic and complete development of maintenance requirements for each system end item, and subassemblies components. 2. Facilitates sorting and combining selected elements of logistics data in support of logistics activities.	Any system programs or projects for which a definition or acquisition phase is applicable or directed and that results in relatively complex nonstandard hardware, facilities or computer program development which will require major logistic support.
Maintenance loading sheet	Correlates maintenance functions and RAS (including frequency of occurrence, time for accomplishment, etc.) to personnel, MGE and spares.	1. Facilitates determination of the quantity of MGE, personnel, and spares required to maintain the system. 2. Provides an input to system effectiveness studies in terms of utilization factors.	All system programs and projects for which definition or acquisition phase is applicable or directed and that will result in or require any one of the following: 1. High launch rates and involving large number of flight vehicles or end items. 2. Large numbers of different end items of a relatively complex and nonstandard nature. 3. Large numbers of end items developed in widely dispersed areas.
MGE utilization sheet	Identifies MGE quantities by specific use location.	1. Facilitates identification of total MGE quantity requirements. 2. Provides input to maintenance loading sheet.	
Personnel utilization sheet	Identifies maintenance personnel effort by specific maintenance location.	1. Facilitates identification of total maintenance personnel requirements. 2. Provides input to maintenance loading sheet.	
Calibration requirements summary	Summarizes equipment calibration requirements at each echelon of calibration.	Provides a convenient summary to define system calibration and measuring standard requirements.	All system programs and projects for which an acquisition phase is applicable or directed that will be turned over to an Air Force using command or that have large number of end items requiring periodic calibration.
Equipment provisioning figure	Defines MGE end item ordering data.	Provides MGE provisioning data to AFLC on an end item basis.	All system programs and projects for which an acquisition phase is applicable or directed and that will result in MGE development which will be procured by AFLC.

Human-Machine Guideline Flow Chart
(Process Subsystem)

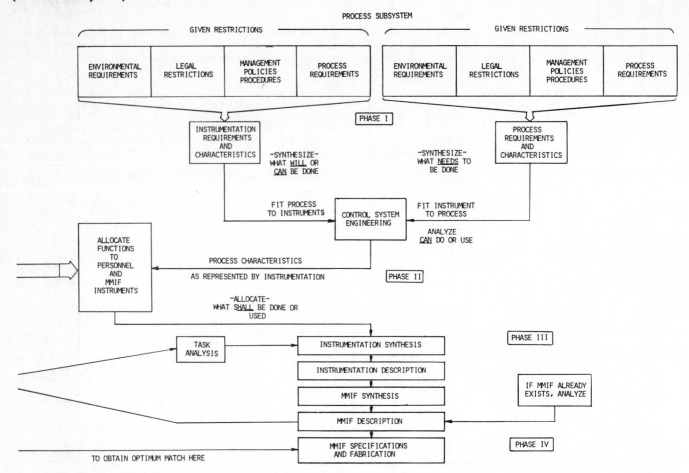

Steps in System Engineering Analysis

Although the following discussion relates primarily to how the military regards system engineering analysis and, more specifically, to how human factors are included in these analyses, the general steps and analysis objectives are equally applicable (although to varying degrees) to any new product or system design development.

In general, system engineering analyses include the following.

1. Mission Requirements Analysis
This involves defining the purpose, objectives, environment, and constraints that any proposed system design is to address. The human factor is an important aspect of this initial set of definitions, especially in terms of the environment in which the system is expected to operate and in terms of certain constraints that the human in the system may place on any proposed design. The environment will relate to physiological limitations of the human; personnel availability and capability may relate to how and where the human can be used.

2. Functional Requirements Analysis
Defining the specific functions that must be performed in order to complete the proposed mission successfully is obviously a critical step

that must be taken before one can decide what to design. Functional requirements are "actions" that have to be accomplished. Before one decides whether actions are accomplished by machines or humans, one should make sure that the functional requirements are clearly defined as actions—not as preconceived notions or ideas about equipment or people.

An important methodology or technique is to develop a graphic model of the function hierarchy in the form of block diagrams; i.e., functional block diagraming is a key part of functional requirements analysis. Functional block diagraming is typically done at several levels, starting with the "top level," where a very gross picture of major functions is shown, each of which will eventually be broken down at several lower levels until a specific critical end-item requirement will emerge, e.g., a piece of equipment, a component, a part, a training aid, a tool, or, most important, a trained operator or maintainer.

3. Function Allocation
As indicated above, to this point, function definitions have been pure action requirements, devoid of decisions about whether hardware or people should perform the functions. The next step, and one in which the human factor becomes extremely important, is to decide which functions can be done more effectively

by machines and which functions can be done more effectively by people. It is critical at this point to have human factors specialists participate in these function-allocating decisions.

4. Detailed Functional Requirements Definition
Although a gross description has already been made for each function defined in step 2, now that allocations have been made (at least preliminarily) as to which functions go to the hardware designer and which go to the human or personnel subsystem designer, each of these specialists proceeds to define his or her respective functional responsibilities in sufficient detail to establish a basis for preliminary design development. Remember that the engineer and the human factors specialist (although each looks at his or her assigned functions from a unique point of view and technical objective) must coordinate with each other in order to ensure an impedance match between machine and human requirements.

5. Preliminary Design Specifications Analysis
As the more detailed requirements are completed for each hardware and/or human function, these are now analyzed in terms of a possible design solution. Engineers are concerned with hardware and facilities designs, and human factors engineers are concerned

primarily with the "design of the personnel subsystem." However, this is the point at which the human factors engineer must also be concerned with what the engineer perceives and how he or she executes hardware and facilities designs, for these designs eventually dictate exactly what people in the system have to do to use the hardware and facilities.

It is in this latter situation that the human factors engineer must provide the engineer with human engineering design principles and criteria so that the design engineer will come up with designs that are compatible with human operator and maintainer capabilities and limitations. During engineering preliminary design, various studies are made to determine which design approach is best, both from an operational and a cost point of view. The human factors engineer should work closely with the engineers and designers during these studies so that design trade-offs will include consideration of operator and maintainer capabilities and limitations. The optimization of designs must reflect both good engineering and good human engineering.

During predesign studies, it may be necessary to perform both engineering and human engineering tests to verify certain assumptions or demonstrate that one design is better than another. It is at this point that mockups are extremely valuable, allowing the designer and the human factors specialist to work together to determine the best compromises between engineering and human engineering objectives.

For large, complex systems, preliminary design typically involves a number of different engineering specialists. These may include, for example, air frame, control systems, avionics, electrical, computer, and other hardware and software specialists or groups. In addition, the typical company usually will have support engineering groups such as integrated logistics, reliability, quality control, factory production, and publications. Each of these groups, along with the primary engineering groups, has primary responsibilities for both the prime system hardware and other end items such as a maintenance and supply logistics plan, a special production facility, development plans for test and other facilities, and finally the human engineering program plan that typically is required by military procurement contracts.

Kept purposely separate from the above is the preliminary design of the engineering test function or functions. It is important to address this independently because of the degree to which human factors should be part of this activity. Although engineering tests are a recognized part of any development program, human engineering testing often is neglected. A final system's performance effectiveness can be measured only in terms of the combined human-machine output. It therefore is critical that the human engineering test requirements be included in the overall system test planning.

A final output of preliminary design, then, is a base-line proposal of what the system is to be like in terms of hardware, software, facilities, support elements, and, most significantly, the operator and maintainer element. Typically, a detailed production plan and operational support plan should be proposed so that

the customer is able to assess not only the basic design concept but also the general costs and schedules required to acquire the system and operate it. The end product is a specification that can be used as the basis for proceeding with detailed design, development, production, test, and delivery of the system. Acquisition or production contracts are written around this specification.

It is perhaps important to pause at this point to illustrate some of the typical analytic procedures that may be used for each of the foregoing analyses. The following examples or samples are meant to show various approaches that different analysts have taken to accomplish the analyses just described. Every analyst tends to feel more at home with a particular style or format for developing information and documenting it for general review by his or her peers, which explains the slight differences among the examples. However, for the sake of clarifying analytic objectives, one should consider the following points:

a. Some sort of numerical system should be created so that each of the analyses can be identified and traced back to initial mission objectives and/or basic functional requirements. Nothing is more embarrassing than to come up with a design for which one cannot find a basic requirement!

b. In addition to verbal descriptions of requirements and/or analytic conclusions, one should attempt to create graphic models that show the flow from one analysis or requirement to another. Block diagraming is a favorite technique, as evidenced by the numerous examples that follow.

c. Functions should be identified by a verb rather than by a noun that describes an equipment or human component.

d. End items should be identified by a noun since these eventually must be applied to a drawing.

e. System schematic diagrams should illustrate both normal and alternative pathways, and all loops must be closed.

f. Drawings, study documentation, test documentation, and other end products should include the identification number of the functions to which they pertain.

How to Decide How Much Analysis and Time Should Be Spent on the Analysis of the Human Factor in Product Development

Obviously, the same amount of analysis is not required for all product designs, either because of the inherent need or because of the amount of time and economic resources that can be made available. The following suggestions may be helpful as a guide in determining the extent to which human engineering analysis should be performed.

1. Need

Need should always be the first consideration. In the case of some products, lack of proper preparatory analysis may lead to costly consequences in terms of redesign, loss of sales potential, and the even more serious possibility of litigation in the event of personal injury

suits. Some guidelines are as follows:

If a proved product design is merely being updated or refined and if its in-use track record indicates reliable operation, good human-machine performance, and no apparent safety problems, extensive human engineering analysis is not usually cost-effective, unless the proposed modifications could possibly introduce hardware- or user-induced failures. In the latter case, human engineering analysis generally is required only on those aspects of the product modification which are new and suspect in terms of potential user misunderstanding, performance failures due to previously learned habit patterns, and/or unique hazards that the modification may introduce.

If the product is brand new but fairly simple in terms of concept and operation, the human engineering analysis requirement is probably minimal. Here it may be necessary to develop only a simple descriptive scenario of how the product will be used. This must be done in order to force the designer to review all the basic operations and maintenance interfaces in order to determine what human engineering design principles should be applied during the design process.

If the product is generally new and relatively complex, a more complete human engineering analysis may be required in order (a) to ensure that the basic concept is compatible with user capabilities and limitations and (b) to derive a more complete set of human engineering design requirements for each subelement that has to be designed. As the system becomes more complex (particularly a new one), there is generally a greater need to perform these human engineering analyses for products that must define manning and training requirements.

Either modified or new products or systems that pose special hazards for users (operators and maintainers) usually require more human engineering analysis—especially analyses of hazards and of failure modes and effects. Typical of these are systems involving excessive energy-producing features; extremes of vibration, noise, or temperature; radiation; toxic contamination; and/or unusual extraterrestrial space conditions. Such systems require the most thorough and complete human engineering analysis because their basic abnormality requires that special precautions be taken concerning the manner in which humans participate in the system and the way they must be protected, not only from the system and the environment but also from their own performance failures.

2. Time Allowance

Time often is a constraining factor in terms of the kind and amount of human engineering analysis effort that can be accomplished before actual product design is commenced. On the other hand, time constraints are too often used as an excuse for not performing the necessary analysis; e.g., arbitrary decisions are made to forgo human engineering analysis in order to meet design schedules (which often are arbitrary and should not be allowed to preclude proper preparation for design). The following are suggested guidelines with respect to providing human engineering analysis time:

A design modification or even a new design that has sufficient operating performance precedence to indicate that there are no significant reasons to doubt that users can operate and maintain the new product as efficiently and safely as the previous or similar one generally does not require extensive preliminary human engineering analyses and therefore much time for analysis. The human engineering concern here is chiefly one of monitoring detailed hardware selection and design on the basis of recognized human engineering design practices, principles, and criteria.

In the case of a design that, although somewhat new, is required as quickly as practicable to meet an urgent operating need, the time allotted to human engineering analysis may be limited; i.e., although one would prefer to take more time, conditions exist that make it mandatory to produce the new or modified product as quickly as possible. Typical examples might include a special weapon needed for use in urgent battle conditions, a device to control traffic at a particularly dangerous intersection, and a safety device to prevent misoperation of an agricultural machine. In such cases, the urgency is established by a higher authority on the basis of the high probability that delaying production will add to already mounting problems within the operational situation. Not only is there insufficient time to perform the desired human engineering analyses, but also the chances are great that the product will already be designed and in use before any reasonably efficient analysis could be completed. Typically this kind of urgency should dictate eliminating human engineering analysis only when the item being designed is for specific, limited application and very limited production runs.

When there is even a shred of doubt about a new design with respect to whether the human's role is compatible with his or her inherent capabilities or about the possible hazards that could exist for the user, time should not be used as an excuse to eliminate desirable human engineering analyses. Rather, the approach should be to examine the product design problem realistically and recommend that sufficient time be provided to do the analyses. This is particularly important for new designs which push the hardware state of the art, which are particularly complex in nature (involving many hardware elements and many human participants), or which appear to require personnel skills that are borderline relative to the capabilities of the expected user population. Generally, if enough time has been provided to do typical systems engineering analyses, there is also enough time to include appropriate human engineering analyses.

3. Level of Effort

The level of human engineering analysis effort (e.g., the number of actual worker-hours) is or should be a function of need versus time—not just cost. At one extreme one should avoid "make-work" analyses costing many worker-hours; at the other extreme one should not expect to derive much benefit from an analysis which was limited to the efforts of too few analysts or which was performed too quickly. Some of the factors to consider in making decisions about the level of effort for human engineering analyses are as follows:

If the design involves a military customer, the level of human engineering analysis effort will invariably be specified by the customer. If this is not clearly spelled out in a request for proposal (RFP), it is best to provide a complete estimate based on the requirements of MIL-H-46855, *Human Engineering Requirements for Military Systems, Equipment and Facilities*. Ordinarily, in reviewing the proposal, the military customer will define how much analysis is required. If the customer does not do so, steps must be taken to clarify this matter before signing the contract.

For designs that do not involve a military or other governmental customer, one should estimate the requirements for a human engineering analysis effort on the basis of the following:

a. If the operating concept is not radically different from that of previous designs demonstrated to be effectively and safely operated by the expected user, the human engineering analysis can probably be limited to a brief identification of operating sequences and/or communications- or information-flow diagrams to help identify significant user-hardware interface requirements.

b. If only one or two new elements are being added to an otherwise older and proved design, analyses should be done only on the new additions. That is, it should not be necessary to perform full-blown, in-depth function allocation, information-flow, or operational sequence analyses on the entire system.

c. On the other hand, if the new design includes untried techniques or procedural concepts, advanced hardware, or applications in unusual environments, the human engineering analysis should be as complete as necessary to ensure that the concept is compatible with the human user's characteristics, capabilities, and limitations. It is wise in this case to make sure that there are enough *trained* human engineering analysts to complete a thorough analysis of all design elements before these are committed to the drawing board.

A good rule of thumb for estimating the amount of effort that should be devoted to human engineering analysis is that such analysis should be roughly 50 percent of the total engineering analysis. This is required because the human engineering analyst generally has to be involved not only in his or her own analysis efforts but also in those of the engineering analysts.

4. Cost

Actual cost predictions for any kind of analysis activity cannot, of course, be made without addressing the specific system or product or without knowing specific manpower and salary rates. However, certain cost factors can be noted that are perhaps useful in approaching the task of costing an analytic program. Some of these are as follows:

Costs will vary depending on the types of analysts required. Human engineering analysis may require a variety of analysts, depending on the type of system under consideration. To start with, the analyses we are dealing with require not only a special kind of person who has had experience and is familiar with the objectives, parameters, and subject matter of human-machine system or product interface design, but also persons with special backgrounds in some instances. For example, in a manned space system, many of the analyses will involve the expertise of an aerospace surgeon or physiologist. Complex information systems may require the expertise of a senior-level psychologist. As a minimum, a qualified human factors engineering analyst is usually a senior-level scientist. It can be seen, then, that the salary level of the people usually required will not be low. This is not to say, however, that lower-salaried individuals cannot learn to be qualified human engineering analysts. In many cases, junior analysts can be used when only minimal analysis is required.

Whether manual or computer analysis techniques will be used also is a determinant in the cost of human engineering analysis. This, of course, depends on the in-house computer facilities and on the desirability of writing programs for the analyses, a question that relates to the extent of the analysis requirement and the desire to minimize the time needed to perform the analyses.

Cost is determined by the size and complexity of the system, i.e., the number of subsystems within the system that involve interactions between equipment elements and people elements. The more elements there are, the greater the number of analyses and thus the greater the costs. As a rule of thumb, one can consider analysis costs as follows:

a. A simple product (a hand tool, a household appliance, an item of furniture, etc.) should require no more than a few hours to analyze the basic functional requirements clearly and determine the interface characteristics and specific design requirements.

b. A more complex product (an equipment console, a test instrument, an electromechanical tool, a fairly simple machine tool, etc.) should require no more than a few days to perform the necessary analyses to establish the above requirements clearly.

c. A multielement system that involves a single operator (e.g., a driver-vehicle system) may require anywhere from a month to several months to perform analyses sufficient to clarify all the human-machine interactions and establish conceptual objectives and design goals.

d. A large, complex, multielement system involving many systems and subsystems and large numbers of operators and support and maintenance personnel (an aerospace system, a ship system, a command and control system, etc.) may require as long as a year or more to complete all the required human engineering analyses. This type of system typically will involve human engineering analyses at different stages, first in the customer's house and later in the contractor's facility. A major reason why this analysis effort is extended is that a great deal of the analyst's time necessarily must be spent in seeking out the necessary information to commence and complete the analyses. And, because of the number of subsystems (each requiring separate analyses), coordination and integration of the various analyses require time over and above that needed for the simple act of constructing block diagrams or flow diagrams or constructing system model graphics and writing descriptive material relative to these graphics. Finally, in almost all complex system analyses, many iterations are required before the final analyses reflect all the intervening trade-offs and decisions that result from system refinement. It is at this level that trained human engineering analysts should always be used.

Military Requirements*

Analysis should include application of human engineering techniques as follows.

1. Defining and Allocating System Functions

The functions that must be performed by the system in achieving its objective should be analyzed. Human engineering principles and criteria should be applied to specify human-equipment performance requirements for system operation, maintenance, and control functions and to allocate system functions to (a) automatic operation and maintenance, (b) manual operation and maintenance, or (c) some combination of these.

2. Information-Flow and Information-Processing Analysis

Analyses should be performed to determine the basic information flow and processing required to accomplish the system objective and should include decisions and operations without reference to any specific machine implementation or level of human involvement.

3. Estimates of Potential Operator and Maintainer Processing Capabilities

Plausible human roles (e.g., those of operator, maintainer, programmer, decision maker, communicator, and monitor) in the system should be identified. Estimates of processing capability in terms of load, accuracy, rate, and time delay should be prepared for each potential operator and maintainer information-processing function. These estimates should be used initially to determine allocation of functions and should later be refined at appropriate times for use in the definition of operator and maintainer information requirements and control, display, and communication requirements. In addition, estimates should be made of the likely effects on these capabilities of

implementation or nonimplementation of human engineering design recommendations. Results from studies in accordance with "studies" requirement may be used as supportive inputs for these estimates.

4. Allocation of Functions

From projected operator and maintainer performance data, cost data, and known constraints, the contractor should conduct analyses and trade-off studies to determine which system functions should be machine-implemented and which should be reserved for the human operator and maintainer.

5. Equipment Identification

Human engineering principles and criteria should be applied along with all other design requirements to identify and select the equipment to be operated, maintained, or controlled by humans. The selected design configuration should reflect human engineering inputs, expressed in quantified or "best-estimate" quantified terms, to satisfy the functional and technical design requirements and to ensure that the equipment will meet the applicable criteria contained in MIL-STD-1472, as well as other human engineering criteria specified by the contract.

6. Analysis of Tasks

The analyses shall provide one of the bases for making design decisions, e.g., determining, to the extent practicable, before hardware fabrication, whether system performance requirements can be met by combinations of anticipated equipment and personnel and assuring that human performance requirements do not exceed human capabilities. These analyses should also be used as basic information for developing preliminary manning levels, equipment procedures, and skill, training, and communications requirements. Those gross tasks identified during human engineering analysis which are related to end items of equipment to be operated or maintained by humans and which require critical human performance, reflect possible unsafe practices, or are subject to promising improvements in operating efficiency should be further analyzed, with the approval of the procuring activity.

7. Analysis of Critical Tasks

Further analysis of critical tasks should identify (a) the information required by the human, including cues for task initiation; (b) the information available to the human; (c) the evaluation process; (d) the decision reached after evaluation; (e) the action taken; (f) the body movements required by the action taken; (g) the work-space envelope for the human required by the action taken; (h) the work space available to the human; (i) the location and condition of the work environment; (j) the frequency and tolerances of the action; (k) the time base; (l) feedback informing the human of the adequacy of his or her actions; (m) the tools and equipment required; (n) the number of personnel required and their specialty and experience; (o) the job aids or reference required; (p) the special hazards involved; (q) the operator interaction where more than one crew member is involved; (r) the operational limits of human performance; and (s) the operational

limits of the machine (state of the art). The analysis should be performed for all affected missions and phases, including degraded modes of operation.

8. Loading Analysis

Individual and crew workload analysis should be performed and compared with performance criteria.

Note: Critical tasks are those which, if not accomplished in accordance with system requirements, will most likely have adverse effects on cost, system reliability, efficiency, effectiveness, or safety.

a. Jeopardized performance of an authorized mission
b. Degradation of the circular error probability (CEP) to an unacceptable level
c. Delay of a mission beyond acceptable time limits
d. Improper operation resulting in a system "no-go," inadvertent weapons firing, or failure to achieve operational readiness alert
e. Exceeding of predicted times for maintenance personnel and maintenance ground equipment to complete maintenance tasks
f. Degradation of system equipment below reliability requirements
g. Damaging of system equipment resulting either in a return to a maintenance facility for major repair or in unacceptable costs, spare requirements, or system downtime
h. A serious compromise of weapon system security
i. Injury to personnel

Checklist for Reviewing System Engineering Analyses

1. Functional Analysis Development

a. Do mission descriptions reflect an accurate interpretation of mission and threat analyses and tactical and nontactical objectives as provided by the personnel data analysis, and are they related to personnel planning data requirements?
b. Are operations and maintenance concepts delineated in sufficient detail to permit application in subsequent task analysis and personnel requirements determination?
c. Do operations and maintenance concepts reflect a correct and complete understanding of personnel capabilities and quantitative limitations?
d. Have functional descriptions and operations system diagrams or other diagramatic presentations of functional relations been updated to reflect progress in system design?
e. Has the installation schedule been modified to reflect any changes?
f. Have detailed equipment descriptions been prepared that provide an accurate and complete description of system equipment as required for personnel planning data purposes? Are they comprehensible to readers who are not engineers?
g. Are equipment descriptions accompanied by illustrative flow diagrams and/or pictures?

*Human Engineering Requirements for Military Systems, Equipment and Facilities, *MIL-H-46588A.

h. Are any items so unclear or undefined that further study and delineation are required?

i. Does a lack of threat analyses or installation schedule data suggest new developments preparation of this portion of the personnel planning data?

2. Human-Machine Assignment

a. Have the human-machine assignments accomplished during contract definition been updated?

b. Have operations systems diagrams or other diagrams been updated? Do the diagrams depict all assignments?

c. Have marginal assignments and problem areas identified during contract definition been resolved?

d. Have additional data sources and techniques employed in the resolutions been clearly identified?

e. Have maintenance assignments been amplified to reflect progress in equipment definition?

f. Do assignments reflect an understanding of personnel capabilities?

g. Do problem areas still exist that need further study? Should these problems be included in the new developments concurrent studies program?

3. External Load Definition

a. Have external load data prepared during contract definition been updated to reflect any additional contractor analyses?

b. Have all input sources been clearly identified?

c. Have levels of activity been clearly related to input sources?

d. Have input sources and levels of activity been clearly related to equipment positions?

e. Do any problem areas or data gaps exist that indicate inclusion in the new developments concurrent study program?

4. Task Analysis Data

a. Have all identified system functions involving humans been included in the analysis?

b. Do task descriptions clearly delineate human inputs and outputs and methods of presenting feedback concerning response adequacy?

c. Are all task inputs clearly identified by data source?

d. Have task sequences been clearly delineated? Do these sequences reflect the results of the external load analyses in that points of peak load are identified in relation to task sequence?

e. Have all tasks requiring rapid, difficult, and/or perceptual-motor combinations of action been fully amplified as required in the personnel planning data specification?

f. Have operations systems diagrams or other diagrammatic depictions of task sequences been prepared in appropriate areas? Are these complete, and do they reflect the external load analysis data?

g. Have task sequences been organized to reflect different system mission and tactical and nontactical objectives as outlined in the functional analysis?

h. Have estimates of task time and task frequency been provided? Has the basis for such estimates been clearly identified?

i. Have performance standards been provided as outlined in the personnel planning data specifications?

j. Are there any areas of task description or analysis that are sufficiently incomplete or inadequately justified so as to require further special study?

5. Position Structure Data

a. Has the guidance contained in the personnel planning data specification been reflected in developing the position structure?

b. Have operations systems diagrams or other diagramatic depictions been prepared to support and illustrate the position structure? Are they in sufficient detail?

c. Do the task groupings reflect the results of external load analyses?

d. Does the position structure reflect the requirements for interaction with other systems within the activity as outlined in the new developments system interface study?

e. Have position descriptions been prepared in accordance with specifications?

6. Knowledge and Skill Requirements

a. Have the knowledge and skill requirements developed during contract definition been updated to reflect the additional detail available concerning system function, operator and maintenance tasks, and position structure?

b. Do the requirements accurately reflect the guidance contained in reference documents?

c. Are the requirements documented in the format and level of detail required by the customer?

d. Are the knowledge and skill requirements supported by identified task and position statements?

Preparing a Human Engineering Plan

Military system development typically calls for submission of a formal human engineering plan as part of a contractor's proposal. Although the content and level of detail may vary, the plan should consider inclusion of the following types of information:

1. Introduction: This is a general statement of the purpose and scope of the proposed effort and includes references to specific regulatory documents to which the plan relates. Key among these documents are MIL-H-46855 (the basic human engineering specification) and MIL-STD-1472 (the basic human engineering standard).

2. Initial guidance meeting: Usually it is required that, upon award of a contract, an initial proposal review meeting be conducted during which the military agency's and the contractor's human engineering personnel will get together to hash out any parts of the proposed human engineering plan that may not be quite to the agency's liking.

3. Personnel and organization: The agency wants to know specifically who in the contractor's organization will work on the program and where these people are in the contractor's organization. It is expected that the human engineering function will be in a position to have an appropriate impact on the entire development program and that it will not be relegated to some obscure level that will be ineffectual.

4. System and task analysis: Human engineering is expected to play an important role in the entire basic system, subsystem, and design analysis effort to ensure that the human factor is given appropriate emphasis at each stage of development.

5. Human engineering design assist: It is important to indicate where and when human engineering will be applied, by whom, and how this effort is to be documented.

6. Mockups: The military expects human engineering to play a significant role in deciding not only what mockups will be constructed but also how they will be used. The military expects mockups to be used as design tools, not merely sales gimmicks.

7. Human engineering tests: The agency wants to know what kinds of specific human engineering tests will be conducted and how and when they will be conducted.

8. Human engineering design verification plan: A separate design verification is required to identify the points at which specific human factors requirements are evaluated. These include all the above as well as special demonstration tests.

9. Documentation and reporting: The military agency will expect to have complete documentation of key phases of the human engineering effort, including monthly reports, special study reports, and all key design reviews. Appropriate engineering drawings are to be attached to all design reviews. If human engineering standards have been compromised for any reason, this is to be reported, including reasons for the compromise.

10. Schedule: A detailed schedule for accomplishing the human engineering effort is to be provided. It should be keyed to the contractor's overall milestone schedule, and the schedule should indicate the level of effort throughout the program.

FUNCTION ALLOCATION

Human-Machine Function Allocation Analysis

Theoretically, decisions regarding how to accomplish each function established by the functional block diagraming and functional requirements analysis should not be made until one is sure that all the necessary functions have been specified and agreed upon. This is to prevent some preconceived idea of how to accomplish a function without serious consideration of several possibilities, particularly whether the function is more appropriate to human or to machine implementation. There are, of course, exceptions, i.e., when past experience has demonstrated that people cannot do certain things and therefore some hardware device *must be used*.

The human-machine function allocation step is extremely important because it allows knowledgeable individuals to contemplate the pros and cons of using a human rather than a machine to accomplish each of the proposed functions. Obviously, one should strive to select the alternative that provides the most cost-effective solution.

Human factors specialists should enter into this phase of the analysis since they are generally more knowledgeable with respect to what humans can or cannot do well, risks that may be too great to expose the human to, severe problems with regard to selecting the right individuals to perform the function, problems associated with training the individual to do an adequate job, and so on. The following guidelines and analytic examples are provided to aid the reader in addressing the function allocation analysis task.

General Considerations for Human versus Machine Function Allocation

ENVIRONMENTAL CONSTRAINTS The human's physiological tolerance to certain operating environments is limited; therefore, one has to make an early decision whether to incur the necessary costs and complexity to protect and support the human under severe environmental conditions (extreme atmospheric pressure, acceleration, temperature, noise, vibration, or radiation and/or potential emergency situations produced by explosive blasts, fire, atmospheric or chemical contamination, etc.).

SENSORY ISOLATION To perform useful tasks within a control environment, humans must be able to receive sensory inputs (information) at levels commensurate with their inherent sensory channel threshold capabilities; e.g., humans can see only so far and hear a signal that is only so soft, and these direct perceptions are easily degraded by various interfering environments. In fact, sensory inputs may be distorted, causing humans to make perceptual errors, i.e., to misinterpret what they see, hear, or feel.

SPEED AND ACCURACY Human response cannot compete with the capacity of a machine in terms of speed and accuracy; thus, functional allocations to humans must be made on the basis of their capacity.

OVERLOAD Humans are fairly limited compared with machines in terms of how much information they can absorb and handle at one time, how many things they can monitor or control at one time, and how effectively they can maintain cognizance of a situation for extended periods of time or under severe physiological and psychological stress conditions.

PHYSICAL STRENGTH Humans are extremely limited compared with machines in terms of how much force they can apply, and for how long.

STORAGE CAPACITY Humans' capacity to store large amounts of information over the long term is extremely great, but their ability to retrieve information quickly is sometimes extremely limited and unreliable; a machine, however, can store almost any amount of data and recall it almost immediately. On the other hand, the machine's capacity to store and retrieve is entirely limited to what is designed into it.

HUMAN-MACHINE PERFORMANCE SURVEILLANCE Humans (compared with machines) are relatively poor "self-monitors" and are easily influenced by emotional factors and by environmental and operational distortions.

INTERPRETATION OF, AND RESPONSE TO, UNEXPECTED EVENTS Humans possess the unique capacity to constantly reevaluate a situation, change their approach, and invent new ideas on the basis of unexpected events and operating conditions. They often can continue with an alternative or less-than-perfect procedure, whereas a machine may quit completely. As noted earlier, a machine does only what it was designed to do; i.e., its capability is limited by the designer's capacity to anticipate all events and conditions of operation.

FATIGUE Humans' capacity and functional capabilities are subject to short- and long-term fatigue effects, whereas machines *can* be designed to be almost fatigue-resistant.

LEARNING Humans generally require some finite learning period to perform a new function. A machine begins its operation immediately and theoretically requires neither initial training nor proficiency refreshment.

COST As long as humans are used properly (within their basic physiological and psychological limits), they often are the least expensive component of a system. Although one obviously must account for the costs of supporting them (housing, pay, medical expenses, etc.), in many cases this has to be done anyway. One must be careful not to try to duplicate human capabilities completely, for such duplication by machine may end up costing much more. A thorough cost comparison is the only sound method for deciding whether to use the human or the machine for given functions. In the case of military systems, the customer must provide complete support for the human. On the other hand, domestic systems and products typically do not require such complete support for the human.

Checklist for Making Decisions Relative to Human-Machine Coupling

In order for humans to complement the capability of a system, they must be coupled with machines in a way that will allow them to utilize their capabilities to the maximum. Thus the following should be considered.

GENERAL CONCEPTUAL PRINCIPLES FOR HUMAN-MACHINE SYSTEM COUPLING

1. Select the sensorimotor link which makes the best use of human capacity, sensitivity, and reliability. Avoid coupling via a particular link merely on the basis of tradition or because it may appear that a particular hardware implementation is less expensive, easier to design, or already available.
2. Choose a coupling approach that maximizes total system effectiveness; do not choose an approach on the basis of whether it is easy or hard to automate a function.
3. Couple humans with machines in such a way that they are not compelled to work at peak limits all or most of the time.
4. Couple humans with machines in such a way that they can recognize or feel that their contribution is meaningful and important. Avoid giving humans machine-serving responsibilities.
5. Couple humans with machines in such a way that information flow and information processing are natural; this minimizes learning time and the probability of confusion or errors.
6. Select coupling methods that do not require extremely precise manipulations; continuous, repetitive movements; frequent, laborious, and lengthy calculations where accuracy is critical; or physical contributions that demand reaching one's upper strength limits.
7. Couple humans with machines as though they might at some time have to assume control (even though the nominal mode may be automatic).
8. Use hardware to aid the human; do not use the human to complement a predetermined hardware concept.

GENERAL PHYSIOLOGICAL CONSIDERATIONS IN HUMAN-MACHINE ALLOCATION ANALYSIS*

Actual physiological hazards to operator health obviously must be considered while determining function allocation to human or machine. However, one must also examine the other boundaries in terms of how (although they may not lead directly to personal injury) conditions might stress operators sufficiently to induce them to make errors and in terms of how these errors might in turn lead to potential system loss and/or eventual injury to operators and others.

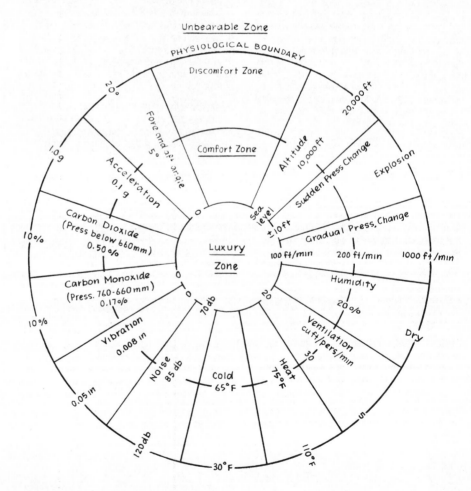

*R. A. McFarland, *Human Factors in Air Transportation: Occupational Health and Safety,* McGraw-Hill Book Company, New York, 1952, p. 705.

TASK ANALYSIS

As the term implies, task analysis involves the examination of what a particular design requires of an operator, maintainer, or product user who must perform a monitoring, control, or maintenance task. The sooner one can examine and define the task implications of a proposed task, the sooner he or she will know whether the design is compatible with the user's capabilities and limitations.

Task analyses and task descriptions serve two main purposes, among others. First, as noted above, the task analysis tells us not only whether the design is feasible but also whether it is optimum from the user's standpoint. Second, the task analysis provides important information for the personnel subsystem specialist who must plan and devise personnel acquisition and training programs for the system.

Preliminary task analyses should be performed as soon as possible during or directly following the initial conceptual phases of a development program; thereafter, they should be regularly refined and updated until the task description becomes a procedural guide.

The accompanying materials are meant to provide guidance in task analysis through the several examples of descriptive materials and documenting formats. It will be observed that various analysts have taken different approaches to documenting their analysis and have used different formats for describing tasks. Obviously, there are no set rules, and therefore these examples merely illustrate the types of information that various analysts felt were necessary to provide useful documentation—for assisting in design review and/or for developing manning and training requirements.

Suggested Information Requirements for Expanded Task Analyses

Item	Description
Operator and maintainer position	Position-type title or specialty rating that it is believed the task should be assigned.
Job operation	Job operation performance description, i.e., operation performed in support of a mission function, usually performed at a single location as a unit of work, in the sense that it has a well-defined beginning and ending.
Duty	Larger units of work under a job operation assigned to an individual in execution of a position. Duties are made up of operationally similar tasks within a given operator or maintainer position. An example might be "carries out overall subsystem test on XX radar."
Task title	A short title that indicates what an individual does (in functional terms).
Task index number	Task-identifying number, usually corresponding with OSD numbering system.
Data sources	Major sources of data useful for later verification (equipment, location, performance criteria, etc.).
Work-area location	Where task is accomplished.
Task description	Includes: Equipment used Displays used Controls used Actions (detailed, sequential description of each task step) Support equipment and aids used Feedback re individual's actions
Type of task	Qualities or characteristics of task (fixed procedure, variable procedure, motor skill, system analysis, circuit analysis, display interpretation, etc.).
Frequency of performance	Once, twice, etc.; hourly; per shift, daily, weekly, monthly.
Performance time	Estimated hours and minutes job or task will take; cite maximum permissible time, etc.
Criticality	Effect of failure to perform on the success or failure of job or mission; potential for personal injury.
Newness	Extent to which the task is new to the using agency.
Other positions	Other tasks whose performance depends on, or interacts with, the performance of other personnel position types (indicate functional interrelations).
Safety factors	Any known or suspected hazards, sources, or preventive requirements.
Tools and equipment	Tools, test equipment, protective gear, reference manuals, etc.
Skills and knowledge	Types and level of skill or knowledge pertinent to the selection and training of individuals for the job or tasks.
Physical characteristics	Special physical requirements necessary for effective task performance (e.g., body size, strength, dexterity, coordination, physiological tolerance to specific and anticipated stresses).

**Example: Analysis to Break Out Tasks
and Subtasks for Human Functions**

Behavior Classification for Task Analysis

Processes	Activities	Specific Behaviors
Perceptual processes	Searching for and receiving information	Detect Inspect Observe Read Receive Scan Survey
	Identifying objects, actions, and events	Discriminate Identify Locate
Mediational processes	Information processing	Categorize Calculate Encode Compute Interpolate Itemize Tabulate Translate
	Problem solving and decision making	Analyze Calculate Choose Compare Compute Estimate Predict Plan
Communication processes		Advise Answer Communicate Direct Indicate Inform Instruct Request Transmit
Motor processes	Simple, discrete tasks	Activate Close Connect Disconnect Join Move Press Set Raise Lower Hold
	Complex, continuous tasks	Adjust Align Regulate Synchronize Track Transport

Source: Adapted from C. Berliner, et al., "Behaviors, Measures and Instruments for Performance Evaluation in Simulated Environments," paper presented at a Symposium and Workshop on Quantification of Human Performance, Albuquerque, N.M., 1964.

Example: Providing Inputs to Design from Task Analysis

TASK CRITICALITY This is not an independent output; rather, it is deduced from the nature of the task and is presented to the engineer with the task description in the form of a note to the task description.

There are three major steps in the derivation of task criticality:

1. *Identify the potential errors which can be made in performance of the task.* This is largely a matter of considering the elements of the task and the perceptual, motor, and decision-making demands imposed on the operator. Thus, in the case of a simple task which involves *(a)* reading a pressure gauge regulating the internal pressure of a rocket and *(b)* stopping a pump at a specified pressure, errors may manifest themselves in two ways:
   a. Failing to stop at the prescribed point
   b. Stopping the pump before the prescribed point
2. *Identify the effect of each potential error on system operation.* In the example above, failing to stop the pump at the prescribed point may result in overpressurization and bursting of the rocket being pressurized. Stopping before the prescribed point will result in underpressurization, which will cause certain sensitive instruments requiring a pressurized atmosphere to function erratically.
3. *Estimate the relative criticality of the potential errors.* Criticality may be scaled in terms of categories such as loss of personnel, destruction of the system, mission failure or abort, mission degradation, and mission delay. In these terms, overpressurization may be more critical than underpressurization, since it may result in explosion of the rocket and destruction of the rocket and launch pad as

well as loss of life, while underpressurization is less critical, since the mission may (though not necessarily will) be degraded.

Pointing out a task as being critical to the engineer "flags" that task as one requiring special consideration in design. Among the solutions which are possible (certainly the list is not exhaustive) are the following:

1. Replacing the human with an automatic means of accomplishing the function if the desired level of correct performance cannot be achieved in any other way
2. Providing means to reduce the probability of error, e.g., assigning a special feedback device to warn the operator when the task is being performed incorrectly
3. Assigning the task to only highly skilled personnel

Task criticality is highly related to specification of task difficulty and error likelihood. Since the engineer thinks in terms of physical effects on the system, it is preferable to flag the task as being critical without indicating that it also has a high difficulty or error-likelihood index. The provision of quantitative indices of a highly precise nature, such as probability of operator error to four figures (e.g., .0013), is not advised, since the engineer cannot interpret the quantitative values in design-relevant terms. A gross categorization of task difficulty, such as the following three-part scale, is as much precision as the engineer can handle in design terms:

1. Simple, routine
2. Somewhat difficult
3. Very difficult

Moreover, it is necessary to apply the above scale only to those few very critical tasks which merit design attention, not to all tasks.

TASK DURATION Task duration should be considered in two ways:

1. As a system requirement, i.e., the time within which the task must be performed

in order to accomplish a given system function
2. As an anticipated human performance capability, i.e., the time within which the operator can actually perform the task

Item 1 is a criterion against which item 2 can be evaluated as satisfying or failing to satisfy system time requirements.

As a system requirement, a task may have to be accomplished in so short a time either that the operator cannot physically perform it or that the probability of the operator's making an error will be substantially increased because of the time loading. In either case, special attention must be drawn to such a task. If the system time requirement is inflexible, it may be necessary to automate the function involved (to eliminate the operator) or else to redesign the manner in which the task can be performed or the equipment is to be operated or maintained.

Task duration is, of course, not critical unless the system's required response time is also critical to the successful accomplishment of the mission. Hence, it is necessary to analyze the mission segment in terms of its time demands before examining any individual task duration. Information required in order to perform task duration analysis will include:

1. System performance time requirements
2. A description of the tasks to be performed by personnel in each mission segment
3. The estimated time required to perform the task

Of these informational requirements, the last is the most difficult to secure because it requires data on the performance time capability of personnel, such as the time required to hook up an umbilical connection. That information can be secured from previous comparable systems in which similar or identical tasks have been timed or from the body of general human performance data in the literature. Neither of these sources is readily available.

TASK DIFFICULTY (ERROR LIKELIHOOD)

The human factors specialist is especially concerned about task difficulty because this, in turn, may lead to a higher error probability, with its attendant effects on mission accomplishment. Task difficulty arises because system requirements are incompatible with and overload the skill capability of the individuals assigned to perform the task.

Task difficulty is not the same as error likelihood. A difficult task need not automatically have a higher error probability if personnel of higher skill are available to compensate for the increased task difficulty. The significance of task difficulty is intensified when the task is also critical to the accomplishment of the mission. Such difficult and critical tasks automatically demand redesign because their attendant error probability cannot be accepted. As in the case of maximum task durations which the operator's performance cannot meet, it may be necessary to automate the performance of the task, relax the accuracy requirement (thus implicitly accepting a higher error probability), or redesign the task to simplify it.

The determination of task difficulty must be made by analyzing the individual task in terms of the inputs which initiate the task (e.g., a verbal message) and the outputs which accomplish the task (e.g., a switch action). The human factors specialist will look for the following characteristics, which may (not necessarily will) indicate an excessively difficult task:

1. The input which initiates the task requires excessively precise visual discriminations or fine motor responses.
2. The operator's response to the initiating inputs must be performed so quickly that he or she has problems keeping up with the initiating inputs.
3. The accuracy demanded of the operator in responding to the initiating inputs is excessive (e.g., heading error must be within 0.5°).
4. The task must be coordinated extremely precisely with other tasks performed by other personnel.
5. The environment in which the task must be performed tends to degrade task performance (e.g., high levels of noise or acceleration).
6. Information from multiple sources (e.g., several displays on a control panel) must be integrated by the operator in order to make a decision.
7. There is less than the desirable amount of information available on the basis of which a decision must be made or an action taken.
8. The task is composed of many subtask elements, the correct performance of which is necessary to task performance, but the amount of feedback provided (knowledge of the correctness or incorrectness of subtask accomplishment) is inadequate.
9. Short-term memory requirements for task performance are excessive (e.g., memory for long sequences or target coordinates).

The design solutions available for reducing task difficulty include the following:

1. Providing additional training or selecting more highly skilled personnel
2. Simplifying the task by such means as combining information sources, providing additional feedback, subdividing the task among several operators, or changing the manner in which the task must be performed
3. Reducing system requirements by accepting a higher error probability, longer response time, etc.

HUMAN ENGINEERING IN DESIGN

The following are excerpts from MIL-H-46855, the human engineering specification used by the U.S. military services. These excerpts pertain specifically to the human engineering effort during design.

Human Engineering in Equipment Detail Design
During detail design of equipment, the human engineering inputs, made in complying with the analysis requirements of previous paragraphs (re: system analysis), as well as other appropriate human engineering inputs, shall be converted into detail equipment design features. Design of the equipment shall meet the applicable criteria of MIL-STD-1472 and other human engineering criteria specified by the contract. Human engineering provisions in the equipment shall be evaluated for adequacy during design reviews. Personnel assigned human engineering responsibilities by the contractor shall participate in design reviews and engineering change proposal reviews of equipment end items to be operated or maintained by man. Human engineering requirements during equipment detail design are specified in the following paragraphs.

Studies, Experiments and Laboratory Tests
—The contractor shall conduct experiments, laboratory tests (including dynamic simulation), and studies required to resolve human engineering and life support problems specific to the system. Human engineering and life support problem areas shall be brought to the attention of the procuring activity, and shall include the estimated effect on the system if the problem is not studied and resolved. These experiments, laboratory tests and studies shall be accomplished in a timely manner, i.e., such that the results may be incorporated in equipment design. The performance of any major study effort shall require approval by the procuring activity.

Mockups and Models—At the earliest practical point in the development program and well before fabrication of system prototypes, full-scale three-dimensional mockups of equipment involving critical human performance (such as an aircrew compartment, maintenance work shelter, or a command control console) shall be constructed. The proposed Human Engineering Program Plan shall specify mockups requiring procuring activity approval and modification to reflect changes. The workmanship shall be no more elaborate than is essential to determine the adequacy of size, shape, arrangement, and panel content of the equipment for use by man. The most inexpensive materials practical shall be used for fabrication. These mockups and models shall provide a basis for resolving access, workspace and related human engineering problems, and incorporating these solutions into system design. In those design areas where equipment involves critical human performance and where human performance measurements are necessary, functional mockups shall be provided, subject to prior approval by the procuring activity. The mockups shall be available for inspection as determined by the procuring activity. Upon approval by the procuring activity, scale models may be substituted for mockups. Disposition of mockups and models, after they have served the purposes of the contract, shall be as directed by the procuring activity.

Dynamic Simulation—Dynamic simulation techniques shall be utilized as a human engineering design tool when necessary for the detail design of equipment requiring critical human performance. Consideration shall be given to use of various models for the human operator, as well as man-in-the-loop simulation. While the simulation equipment is intended for use as a design tool, its potential relationship to, or use as, training equipment shall be considered in any plan for dynamic simulation.

Equipment Detail Design Drawings—Human engineering principles and criteria shall be applied to equipment drawings during detail design to assure that the equipment can be efficiently, reliably and safely operated and maintained. The following drawings are included: panel layout drawings, communication system drawings, overall layout drawings, control drawings and other drawings depicting equipment important to system operation and maintenance by human operators. The approval of these drawings by the contractor shall signify that human engineering requirements are incorporated thereon and that the design complies with applicable criteria of MIL-STD-1472 and other human engineering criteria specified by the contract.

Work Environment, Crew Stations and Facilities Design—Human engineering principles and criteria shall be applied to detail design of work environments, crew stations and facilities to be used by man in the system. The approval of drawings, specifications and other documentation of work environment, crew stations and facilities by the contractor shall signify that human engineering requirements are incorporated thereon and that the design complies with applicable criteria of MIL-STD-1472 and other human engineering criteria specified by the contract. Design of work environment, crew stations and facilities which affect human performance, under normal, unusual and emergency conditions, shall consider at least the following where applicable:

a. Atmospheric conditions, such as composition, volume, pressure and control for decompression, temperature, humidity and air flow.
b. Weather and climate aspects, such as hail, snow, mud, arctic, desert and tropic conditions.
c. Range of accelerative forces, positive and negative, including linear, angular and radial.
d. Acoustic noise (steady state and impulse), vibration, and impact forces.
e. Provision for human performance during weightlessness.
f. Provision for minimizing disorientation.
g. Adequate space for man, his movement, and his equipment.
h. Adequate physical, visual, and auditory links between men and men, and men and their equipment, including eye position in relation to display surfaces, control and external visual areas.
i. Safe and efficient walkways, stairways, platforms and inclines.
j. Provisions for minimizing psycho/physiological stresses.
k. Provisions to minimize physical or emotional fatigue, or fatigue due to work-rest cycles.
l. Effects of clothing and personal equipment, such as full and partial pressure suits, fuel handler suits, body armor, polar clothing, and temperature regulated clothing.
m. Equipment handling provisions, including remote handling provisions and tools when material and environment require them.
n. Protection from chemical, biological, toxicological, radiological, electrical and electromagnetic hazards.
o. Optimum illumination commensurate with anticipated visual tasks.
p. Sustenance and storage requirements (i.e., oxygen, water and food), and provision for refuse management.
q. Crew safety protective restraints (shoulder, lap and leg restraint systems, inertia reels and similar items) in relation to mission phase and control and display utilization.

Human Engineering in Performance and Design Specifications—The provisions of performance and design specifications, prepared by the contractor, shall conform to applicable human engineering criteria of MIL-STD-1472 and other human engineering specified by the contract.

Equipment Procedure Development—Based upon the human performance functions and tasks identified by human engineering analyses, the contractor shall apply human engineering principles and criteria to the development of procedures for operating, maintaining or otherwise using the system equipment. This effort shall be accomplished to assure that the human functions and tasks identified through human engineering analysis are organized and sequenced for efficiency, safety and reliability and to assure that the results of this effort shall be reflected in the development of training and technical publications. The approval of these publications by the contractor shall signify that the human engineering requirements are incorporated therein.

Human Engineering in Test and Evaluation—The contractor shall establish and conduct a test and evaluation program to: (1) assure fulfillment of applicable requirements herein; (2) demonstrate conformance of system, equipment and facility design to human engineering design criteria; (3) confirm compliance with performance requirements where man is a performance determinant; (4) secure quantitative measures of system performance which are a function of man-machine interaction; and (5) determine whether undesirable design or procedural features have been introduced. (The fact that these functions may occur at various stages in system or equipment development shall not preclude final human engineering verification of the complete system. Both operator and maintenance tasks shall be performed as described in approved test plans during the final system test.)

Planning—Human engineering testing shall be incorporated into the test and evaluation program and shall be integrated into engineering design tests, contractor demonstrations, R & D acceptance tests and other major development tests. Compliance with human engineering requirements shall be tested as early as possible. Human engineering findings from early testing shall be used in planning and conducting later tests.

Implementation—The human engineering test and evaluation program,

HUMAN ENGINEERING
Suggested Steps for Product/System Design

164

contained in approved test plans, shall be implemented by the contractor. Test documentation (e.g., checklists, data sheets, questionnaires, schedules, operating procedures, test procedures) shall be available at the test site. Human engineering portions of all tests shall include, where applicable, the following:

a. A simulation (or actual conduct where possible) of mission or work cycle.

b. Tests in which human participation is critical with respect to speed, accuracy, reliability or cost.

c. A representative sample of non-critical scheduled and unscheduled maintenance tasks.

d. Proposed job aids.

e. Utilization of personnel who are representative of the range of the intended military user population in terms of skills, size, and strength and wearing suitable military garments and equipment which are appropriate to the tasks, and approved by the procuring activity.

f. Collection of task performance data.

g. Identification of discrepancies between required and obtained task performance.

h. Criteria for the acceptable performance of the test.

Failure Analysis—All failures occurring during, or as a result of test and evaluation shall be subjected to a human engineering review to differentiate between failures due to equipment alone, man-equipment incompatibilities and those due to human error. The procuring activity shall be notified of design deficiencies which contribute to human error.

Cognizance and Coordination—The human engineering program shall be coordinated with maintainability, system safety, reliability, personnel, training and other related programs, and shall be integrated into the total system program. The human engineering portion of any analysis, design or test and evaluation program shall be conducted under the direct cognizance of personnel assigned human engineering responsibility by the contractor.

HUMAN ENGINEERING SURVEILLANCE DURING THE DESIGN PROCESS

It does little good to express a resolve to think about the human factor during the design process or to pass out human engineering design handbooks. Experience has shown that the designer has too many other factors to think about and thus will probably forget about the human element until decisions are made and constraints are designed into the system, equipment, or product. Thus, whether a human factors specialist is assigned as a "watchdog" or whether some member of the design staff is given the human engineering responsibilities, constant surveillance is required. The following suggestions are offered to help designers plan a reasonably effective program for monitoring the human engineering aspects of design:

1. Identify specific design efforts which are expected to be accomplished during the program or project and which appear to have potential human operator, maintainer, or user interface features (they will contain displays, controls, handles, or fasteners; they must be manually lifted, moved, or carried; they probably will produce noise, vibration, heat, radiation, or toxic fumes; etc.).

2. Identify the designer who will be responsible for each design and determine the schedule for that design to begin.

3. Make initial contact with each designer and review the human factors concerns with each one. Discuss human engineering objectives and concepts. Set up a schedule for repeat contacts at regular intervals to review the progress of the design and offer to consult any time the designer has a question about human factors. Above all, be available and point out that you are there to help, not criticize.

4. Review all top-level drawings and from these identify lower-level drawings that have pertinent human engineering features. Review these quickly so that the drawing review cycle is not delayed. Prepare a written evaluation with recommendations. Then discuss the recommendations directly with the designer before passing the drawing on to the next person. Document the results of the discussion, indicating whether your recommendations are accepted or rejected. If rejected, document the reason.[18]

5. Participate in or observe pertinent engineering mockup evaluations and laboratory tests.

6. Participate in vendor product evaluations and assist in component selection.

7. Participate in final drawing-release review, insisting on sign-off responsibility before drawings can be released to manufacturing.

8. Participate in customer design reviews.

9. Monitor factory production and assembly of pertinent parts of the system. Last-minute "glitches" often appear in the assembly shop, especially with respect to maintenance-type tasks (inaccessibility of fasteners, connection reversal features, etc.).

10. Participate in prototype testing. Although test engineers are skilled at detecting, analyzing, and reporting hardware failures, they generally do not recognize design-induced human failures that can be caught during this type of testing.

SUGGESTED STEPS TO ENSURE PROPER HUMAN ENGINEERING IN PRODUCT AND SYSTEM DESIGN

1. Review the mission, purpose, and use scenario for the proposed product or system until you fully understand everything possible about the basis for the product or system and the conditions under which it is to be used.

2. Where practical, engage the services of a qualified human engineering specialist to assist and advise throughout the development of the proposed design and its final production. When this is impractical, consider the following steps.

3. Acquire and/or develop a "human engineering design checklist" to be used throughout the design development and production cycle. Although general checklist examples can be found (some of which are provided in this chapter), take time to create a checklist that is tailored to the nature of your proposed design. This checklist typically requires modification as the design development progresses since, as one comes to know more about the design features, additional user-hardware interface features generate additional human engineering questions that must be addressed and monitored throughout the design cycle.

4. Acquire appropriate human engineering design guides and references. If a qualified human factors engineering specialist is available, he or she probably will already have these. If not, obtain copies of at least the following:

a. MIL-STD-1472 *Human Engineering Design Criteria for Military Systems, Equipment and Facilities* (for products or systems oriented toward the military).

b. Van Cott and Kincaide, *Human Engineering Guide to Equipment Design*, Government Printing Office, Washington.

c. J. F. Parker and V. R. West (eds.), 2d ed., *Bioastronautics Data Book*, National Aeronautics and Space Administration, Washington, 1973.

5. Review initial design concept ideas (either your own or those of subordinates who will be responsible for each portion of the design) to make sure that all potential user interface aspects of the proposed *design concept* have been properly identified and considered in each idea. The purpose of this step is to avoid establishing constraints to good human engineering practice.

6. Using the above human engineering checklist, monitor each design activity as it progresses from preliminary through detailed design steps, making sure that the human factors are kept constantly in mind and that good human engineering practice, principles, and criteria are being considered during each step of the design process. Keep a running record of whether human engineering principles and criteria are being applied and, if compromises are being made, of how and why these are being introduced in the design. This record may be extremely important, for example, if the military customer requires evidence that you have not ignored human

factors and/or justification for abrogating some human engineering principle or if a customer brings suit against you, and you need evidence to prove that you did the best you could to prevent misuse of the product, i.e., that you took all practicable means to minimize the probability of misuse or potential hazard to the customer.

7. Use mockups to "test" the efficacy of all user-hardware interface designs, using "real people" as subjects. Examine and evaluate the mockup-operator interfaces in terms of human performance efficiency (time, error, inconvenience, comfort, inadvertent hazard potentials, etc.). Record these observations and seek appropriate design modifications. Modify the mockup and reevaluate.

8. Perform experiments when necessary to establish design criteria where previously cited reference guides do not provide adequate information for design decisions. This may require development of special, dynamic, real-time simulations of procedural and environmental conditions.

9. Whenever possible, fabricate a hardware prototype and evaluate this under real-life conditions (using typical expected user subjects as well as expert operators). Obtain quantitative performance measures of the total user-hardware operation to prove that the combined human-machine operation is satisfactory. Make sure that all deficiencies are fed back into the design cycle and that appropriate design modifications are made. Retest if necessary.

10. Critically review final production drawings to make sure they are correct before being released for final production fabrication and assembly.

11. Identify critical areas where lack of proper production quality control might result in poor user-hardware interface results because of "sloppy" manufacturing or assembly procedures. Take steps to establish necessary procedural control and inspection to preclude mistakes in the factory.

12. Perform production (field) tests of production hardware before approving it for final delivery to the customer. These tests should include "man-in-the-loop" exercising, not merely visual inspection and/or test of the hardware and software components alone.

Typical Methods for Human Engineering Evaluation of System and Component Design

Method	Purpose	Typical Use
Worker and consumer interrogation	To define general product user needs	When a system or product is to be redesigned or new ones are to be developed, it is desirable to ask people who are using a similar product or system how they feel about it in terms of adequacy to do the job, ease with which the job can be done, and/or special problems that occur with the present hardware, facility, or tools.
Human-machine operational observation	To define general, dynamic, and environmental factors associated with product use	Although interrogation of the user may include demonstration, it is advantageous to observe an operation covertly over an extended period of time in order to have a more objective assessment of problems the operator may have.
Personal operation experience	To provide the designer with a personal "feel" for the problems a worker may have identified or may have observed from a distance	When practicable, it is recommended that the designer actually operate or use a product similar to the one he or she intends to redesign and/or create in order to experience at firsthand the problems or needs which a worker has pointed out or which the designer may have observed from a distance.
Time and motion study	To measure task performance against a time base and thus identify critical product or task conditions	Especially useful for systems in which worker output is a significant factor, i.e., in which increase in rate of production is a primary objective of the system. The method provides quantitative information that helps establish priorities for design and procedural improvement, either through redesign and/or through new design.
Preliminary design review	To evaluate the general, overall probability that a design will meet original operational objectives	Preliminary design drawings provide the first general picture of what a system or product will look like and how it will probably work. It is especially important at this stage to examine the product-user relationships to make sure that these are compatible with operator, maintainer, and user characteristics and limitations, prior to spending too much time with detailed design efforts.
Mockup evaluation	To evaluate both general and specific three-dimensional relationships between various elements of a proposed system and/or specific relationships between a specific part of the product or system and its user	Miniature-scale mockups provide an overall view of a system that could not be viewed effectively in full scale. Full-scale mockups provide one-to-one evaluation of direct product-operator physical relationships and quasi-operating conditions.

(continued)

Typical Methods for Human Engineering Evaluation of System and Component Design *(continued)*

Method	Purpose	Typical Use
Simulation	To investigate and verify dynamic aspects of a proposed design prior to final commitment. To provide an opportunity to explore alternatives both during predesign and during detailed design	A proposed design concept may be examined and/or tested at several levels prior to committing it to preparation of preliminary or detailed design. It is important to include the human user during such simulations since the ultimate effectiveness of the product includes the typical user input, both intentional and unintentional. Simulation should be in "real time" and should include representative environmental, dynamic, and personnel elements, depending upon the stage of design at which the simulation effort is being performed. Evaluative methods should include quantitative measures of system or product-operator performance.
Detailed design drawing review	To verify human-machine and product-user compliance with good human engineering practice	Methodical analysis of detailed design drawings prior to their release to production is vital to prevent typical oversights of human-machine interface problems. This is critical because of the probability that changes may have been made since original human engineering inputs were made at the predesign stage.
Prototype testing	To verify the operating and maintenance effectiveness of the system or product under actual operating conditions	Whenever practicable, a prototype of the product or system should be tested prior to committing it to production. Such tests should include use of operator, maintainer, and user personnel who are representative of the final user population. Although initial tests may be made by special test personnel in order to ensure safety of operation, the ultimate question is whether typical users can and will operate the system or use the product as planned. Quantitative measurement of human-machine performance should be accomplished whenever the complexity, integrity, or safety of the system is critical.
Field evaluation	To provide a continuing assessment of system or product effectiveness under typical use and environmental conditions, in order that service requests can be effectively met, design modification requirements can be defined, and/or new concepts can be anticipated	Systematic follow-up using formal data and information retrieval methods provides a means for a manufacturer to maintain adequate customer relations and to develop insight useful for expanding the product's capability. To provide future human engineering guidelines, useful human factors data should be collected at the same time that pure hardware data are being collected.
Human factors experiments	To support current and/or future design requirements	Special human factors experiments may be required at various times, either during a product development and/or independently. Cognizance of applicable human factors research information is required at the beginning of any development program to assist in defining the human role in system operation. When research data do not exist, special studies may be required to define human capabilities with respect to a specific, proposed application. Whether to perform such research depends on the criticality of the question and/or the time available to perform the research. Such research should have as broad an application as possible, as long as it provides the necessary answers to the immediate problem; i.e., when the answer is too specific, the cost is increased over the long run, since the research may have to be repeated for the next program.

PRODUCT EVALUATION CHECKLIST

The following is a checklist for evaluating whether a product is compatible with user behavioral expectancies:

1. Is it obvious what the product is and what it is to be used for?
2. Is it obvious how the product is to be used?
3. Is the product simple to use, i.e., is it simple to prepare for use, to begin to use or operate, to continue to operate, to stop using or operating, and to place in a stored or nonoperational condition?
4. Are there hidden hazards to the operator or user, i.e., hazards that cannot be readily observed when looking at the product?
5. Will use of the product create hazards for others not involved in its use?
6. Could the product be operated and misused by someone who should not be using it, such as a child?
7. Could the product be used for some purpose for which it was not intended?
8. Should the product be used only under certain conditions, and is this immediately obvious to the potential user?
9. Are potential product failures or conditions under which the product should not be operated or used identified for the user in time to avoid misuse?
10. Would use by someone who is fatigued, under the influence of drugs or alcohol, or otherwise incapacitated be hazardous to the user or others—and if so, does the design provide appropriate use-prevention features?
11. Are there potential hazards associated with the product when it is not being used, when it is stored, or when it is otherwise unattended—and if so, does the design provide appropriate use-prevention features?
12. Is it critical that the product be serviced or maintained in a specific manner—and if so, are the servicing requirements made clear to the user?
13. Is the product prone to damage or failure—and if so, does it contain a fail-safe mechanism?
14. Is the product easy and convenient to service and maintain without special training—and if not, is this clear to the user?
15. Can the product be used when it is not functioning properly without creating a hazard to the user?
16. Is the product, in any way, outside the normal user's physical, mental, or normal behavioral capacities (e.g., in terms of size, weight, or operation)?

CHECKLISTS

Preliminary Human Engineering Checklist for Initial Hardware Selection and Design Analysis

The checklist below provides a starting point for gross evaluation, but it should be expanded for detailed design evaluation. Evaluation designations are as follows: S—satisfactory; C—compromise but acceptable; and U—unsatisfactory. Various other designations are also possible to include evaluative indications of temporary status.

A. EQUIPMENT OPERATION | S | C | U
1. CONSOLE SHAPE/SIZE
   a. Desk height, area
   b. Control reach
   c. Display view
   d. Body, limb clearance

2. PANEL LOCATION
   a. Frequency of use
   b. Sequence of use
   c. Emergency response
   d. Multi-operator use

3. PANEL LAYOUT
   a. Functional grouping
   b. Sequential organization
   c. Identification
   d. Spacing for clearance

4. DISPLAYS
   a. Functional compatibility for intended purposes
   b. Intelligibility of information content
   c. Control interaction
   d. Legibility; figures, pointers, scales
   e. Visibility; illumination, parallax
   f. Location
   g. Identification

5. CONTROLS
   a. Functional compatibility for intended purpose
   b. Location, motion, excursion and force
   c. Display interaction
   d. Spacing, clearance, size
   e. Identification

B. ASSEMBLY - SERVICES - MAINTENANCE
1. INSTALLATION, SERVICE, & MAINT ACCESSIBILITY
   a. Location, size of openings
   b. Covers, fastening/removal
   c. Identification

2. EQUIPMENT HANDLING/TRANSPORT
   a. Size/shape/weight/balance
   b. Handling clearance
   c. Handling aids
   d. Instructions/labels/warnings

3. CHASSIS LAYOUT, PACKAGING | S | C | U
   a. Ease of handling.
   b. Access to components for test, (component) replacement
   c. Identification
   d. Hazard/Damage protection

4. CABLES/LINES/CONNECTIONS
   a. Ease and security of assembly-disassembly
   b. Connection error
   c. Identification
   d. Access, test, trouble-shooting replacement

C. SYSTEM SAFETY
1. PERSONNEL HAZARDS
   a. Shock
   b. Burns: direct, chemical
   c. Hearing damage
   d. Tripping/falling
   e. Pinching
   f. Cutting
   g. Bumping

2. EQUIPMENT DAMAGE
   a. Electrical overload, short, ground
   b. Mechanical overload, strip, bend, rupture, break
   c. Explosion/fire

D. GENERAL
1. LABELS/MARKING
   a. Intelligibility
   b. Legibility
   c. Location, spacing
   d. Permanence

2. EQUIPMENT FINISH
   a. Color
   b. Texture
   c. Reflectivity

3. STORAGE
   a. Location
   b. Volume
   c. Material accessibility, security

4. WORK AREA ILLUMINATION
   a. Light level; range, control
   b. Distribution, contrast
   c. Color

**Human Error Potential as a Basis for
Design Analysis***

Error Prevention through Good Design

Causes of Primary Errors	Preventive Measures (Taken by Designer or Methods Engineer)
1. Improvising procedures that are lacking in the field	1. Provide adequate instructions.
2. Following prescribed but incorrect procedures	2. Ensure that procedures are correct.
3. Failure to follow prescribed procedures	3. Ensure that procedures are not too lengthy, too fast, or too slow for good performance, and are not hazardous or awkward.
4. Lack of adequate planning for error or unusual conditions.	4. Provide backout or emergency procedures in instructions.
5. Lack of understanding of procedures.	5. Ensure that instructions are easy to understand.
6. Lack of awareness of hazards.	6. Provide warnings, cautions, or explanations in instructions.
7. Untimely activation of equipment.	7. Provide interlocks or timer lockouts. Provide warning or caution notes against activating equipment unless disconnected or disengaged from load, or other damaging conditions.
8. Errors of judgment, especially during periods of stress.	8. Minimize requirements for making hurried judgments, especially at critical times, through programmed contingency measures.
9. Critical components installed incorrectly.	9. Provide designs permitting such components to be installed only in the proper ways. Use asymetric configurations on mechanical equipment or electrical connectors; use female or male threads or different-sized connections on critical valves, filters, or other components in which direction of flow is important.
10. Exceeding prescribed limitations on load, speed, or other parameter.	10. Provide governors and other parameter limiters. Provide warnings on: exceeding limitations, inadequate strength of stressed parts, and use of excessive mechanical leverage.
11. Lack of suitable tools or equipment.	11. Ensure that need for special tools or equipment is minimized; develop and provide those that are necessary; stress their need in instructions.
12. Interference with normal habits.	12. Ensure that recognition and activation patterns are in accordance with usual practices and expectancies.
13. Lack of data on which to make correct or timely decisions.	13. Ensure that response time is adequate for corrective action; if not, provide automatic corrective devices.
14. Hampered activities because of interference between personnel.	14. Ensure that space is adequate to perform required activities simultaneously.
15. Inability to concentrate because of unsafe conditions or equipment.	15. Ensure that personnel must not work close to unguarded moving parts, hot surfaces, sharp edges, or other dangers.
16. Error or delay in use of controls.	16. Avoid proximity, interference, awkward location, or similarity of critical controls. Locate control close to readout. Locate readouts above control so hand or arm making adjustment does not block out readout instrument. Ensure that controls are labeled prominently for easy understanding.
17. Error or delay in reading instruments.	17. Ensure that instruments are labeled and designed for easy understanding; do not require reader to turn head or move body; and that visibility problems due to glare or lack of light, legibility, viewing angle, contrast, or reflections are avoided. Provide direct readings of specific parameters so operator does not have to interpret.
18. Inadvertent activation of controls.	18. For critical functions provide controls that cannot be activated inadvertently: use torque types instead of push buttons. Provide guards over critical switches.
19. Controls activated in wrong order.	19. Place functional controls in sequence in which they are to be used. Provide interlocks where sequences are critical.
20. Control settings by operator not precise enough.	20. Provide controls that permit making settings or adjustments without need for extremely fine movements. Use click type controls.

*Willie Hammer, *Handbook of System and Product Safety*, Prentice-Hall, Inc., Englewood Cliffs, N.J., 1972, p. 72.

(continued)

Error Prevention through Good Design *(continued)*

Causes of Primary Errors	Preventive Measures (Taken by Designer or Methods Engineer)
21. Controls broken by excessive force.	21. Ensure that controls are adequate to withstand maximum stress an operator could apply. Provide warning and caution notes for those devices that could be overstressed.
22. Failure to take action at proper time because of faulty instruments.	22. Provide procedures to calibrate instruments periodically, or provide the means to ensure during operation that they are working correctly.
23. Confusion in reading critical instruments because of instrument clutter.	23. Make critical instruments most prominent or locate in easiest to read area.
24. Failure to note critical indication.	24. Provide suitable auditory or visual warning device that will attract operator's attention to problem.
25. Involuntary reaction or inability to perform properly because of pain due to burns, electrical shock, puncture wound, or impact.	25. Insulate or guard against hot surfaces, "live" electrical conductors, sharp objects, and hard surfaces.
26. Fatigue.	26. Avoid placing on operator severe and tiring physical and mental requirements such as loads, concentration times, vibration, personal stress, and awkward positions.
27. Vibration and noise cause irritation and inability to read meters and settings or to operate controls.	27. Provide vibration isolators or noise elimination devices.
28. Irritation and loss of effectiveness due to high temperature and humidity.	28. Provide environmental control. Prevent entrance or generation of heat or moisture from external sources or from internal equipment or processes.
29. Loss of effectiveness due to lack of oxygen, or to presence of toxic gas, airborne particulate matter, or odors.	29. Prevent generation or entrance of contaminants into the occupied space. Provide suitable life support equipment. Avoid presence near occupied areas of lines or equipment containing hazardous gases or liquids.
30. Degradation of capabilities due to extremely low temperatures.	30. Ensure that design provides for adequate heating or insulation, protective shelter, equipment, or clothing.
31. Fixation or hypnosis.	36. Avoid procedures or designs that require visual concentrations for long periods of time. Avoid humming equipment. Provide alternate reference points. Provide procedures to relieve monotony.
32. Disorientation or vertigo.	32. Provide adequate reference points or means to maintain orientation.
33. Slipping and falling.	33. Incorporate friction surfaces or devices, guard rails, access hole covers on floor openings, or protective harness in designs.
34. Inattention.	34. Avoid long intervals between procedural steps. Provide female voice on audio devices to attract attention. Provide bright, colorful, and pleasant work areas.

General Checklist for Safety Achievement via Appropriate Analysis and Design*

Special Safety Considerations

Because of the special nature of some products, it is sometimes necessary to consider safety from more than one point of view. These problems are best explained through examples:

A hand-held weapon will always have certain inherent hazards because of the fact that the device is designed purposely to fire a projectile. Experience has shown that such a weapon should always have some type of safety mechanism to help prevent unplanned firings. The position and specific articulation of the safety feature should be such that the device is convenient to use after the weapon is placed in the firing position; this is necessary in order to minimize the possibility of discharging the weapon while the user is carrying it or is mounting it to the firing position. However, the position and articulation of the safety device should not be such that it can be activated by inadvertent brushing of the device as the weapon is being carried or lifted to firing position or at any time the weapon is being serviced. In addition, since weapons often may be found by children, the safety device should be designed so that a child *cannot* operate and thus remove the safety feature.

A power tool or implement should always have a safety feature to prevent inadvertent operation, and the safety device must be convenient to operate once the tool is in the operating position. Designing and locating the safety device require careful consideration of *all* possible points in the handling cycle in order to make sure that careless handling will not also deactivate the device. In addition, there should be an automatic deactivation of power when the tool is set down temporarily so that no one else could inadvertently pick up the tool and accidentally start the motor, thinking that it had been turned off.

Child-proof bottle caps are generally required by law on all medicine bottles. Remember, however, that some medicines must be quickly accessible to certain persons (e.g., an elderly heart patient), and it is therefore vital that the individual needing to open the bottle can do so easily and rapidly. Confusing safety caps or ones that require considerable finger strength may make the medicine inaccessible to the person who needs it as quickly as possible.

Safety belts obviously serve a basic safety need. However, in addition to providing security for the passenger during a crash, they must also be easy to remove in an emergency—either by the individual wearer or by a rescuer, in the event the wearer may be unconscious. One must consider both the location and the method of operation of the release device in terms of how difficult it might be to find and operate under abnormal conditions, such as in the dark, when

*Willie Hammer, *Handbook of System and Product Safety*, Prentice-Hall, Inc., Englewood Cliffs, N.J., 1972, p. 253.

Safety Measures

Accident Prevention	Damage Minimization and Control
1. Hazard elimination	1. Isolation
2. Hazard level limitation	*a.* Distance
a. Intrinsic safety	*b.* Energy absorption
b. Limit-level sensing control	*c.* Deflection
c. Continuous monitor and automatic control	*d.* Containment
3. Lockouts, lockins, and interlocks	Hazard
a. Isolation	Operation
b. Lockouts and lockins	Personnel
c. Interlocks	Material
4. Fail safe designs	Critical equipment
a. Fail passive	2. Personal protective equipment
b. Fail active	*a.* Programmed dangerous operation
c. Fail operational	*b.* Investigations and corrections
5. Failure minimization	*c.* Emergencies
a. Monitoring	3. Minor loss acceptance
b. Warning	4. Escape and survival
c. Safety factors and margins	*a.* Point of no-return warning
d. Failure rate reduction	*b.* Crashworthiness designs
Derating	*c.* Escape and survival equipment
Timed replacements	*d.* Escape and survival procedures
Screening	5. Rescue
Redundancy	*a.* Procedures
6. Backout and recovery	*b.* Equipment
a. Normal sequence restoration	
b. Aborting entire operation	
c. Inactivating only malfunctioning equipment	
Automatic	
Manual	

the vehicle is not upright, and when it is not possible to open a door.

Standardization plays an important part in safety in a variety of situations and should be kept in mind during any hazards analysis. The following considerations are important:

1. Critical safety features (door handles, safety latches, light switches, etc.) should always be illuminated so that they can be located and operated at night.
2. Standard safety devices (light switches, door handles, restraint system release handles, fire extinguishers, first-aid boxes, etc.) should be located where people expect to find them.
3. Safety signs and signals should be located where people normally will be looking, and they must not have the potential of being hidden by an object placed in front of them or by individuals standing in front of them at a critical moment.
4. Safety devices should operate in a standardized manner; i.e., one device should not have a switch or handle that moves in one direction, while another, similar device has a handle that moves in the opposite direction. The split second required to correct an action may result in injury.
5. A safety device should never be used that may malfunction because of loss of electrical power. A safety system should provide maximum security throughout any anticipated operating cycle; i.e., avoid a system which may provide an initial safety condition but which, because of the nature of the device, expends its safety capability in "one shot" and thus is unavailable for continuing hazardous conditions. The automobile

air bag is a case in point. It is effective only on the first impact, deflating shortly thereafter. Thus, if the crash scenario continues, the passenger no longer has any safety protection.

Toys present a particularly difficult safety problem in that they are subject to breakage by the child and thus present hazards after they are broken as well as before.

In order to resolve some of the special problems exemplified by the above, one should systematically analyze both the normal use and potential abnormal or misuse modes of the particular product. In doing this, keep in mind the following critical human characteristics:

1. The users' level of intelligence (both intended and unintended users)
2. The users' experience with this particular device
3. The various environmental conditions under which the device may be used
4. The physical characteristics of the individual user population (both intended and unintended users), i.e., size, strength, mobility, dexterity, visual and auditory capacity, reaction time, and possible handicaps

To correct safety deficiencies, consider the following:

1. Remove the hazard altogether if possible.
2. If that is not possible, try to minimize the potential effects of the hazard.
3. Provide a barrier between the hazard and the user.
4. Warn the user of the potential hazard.

Finally, do not simply assume that your solution to a safety problem works. Test it.

MOCKUPS

Use of Mockups for Human Engineering Test and Evaluation

Mockups of products are widely used both as a design tool and for marketing purposes in industry. Unfortunately, management perceives the value of the latter use more readily than the value of the former. This section attempts to champion the use of mockups for purposes of design evaluation, especially for validating design features that involve user interfaces.

Types of Mockups

Two distinct types of mockups, each used for different purposes, are of interest. The first is the miniature-scale mockup, which is valuable in examining three-dimensional relationships between several fairly large features that would be difficult to view all at once if they were demonstrated in full scale. The second is the full-scale mockup, which is valuable in examining the immediate three-dimensional relationships between the system being mocked up and an operator or maintainer, in terms of space, reach, visibility, and convenience.

Levels of Mockup Complexity and Detail

Although many people think of a mockup as a fairly substantial structure simulating the three-dimensional attributes of a work station or space layout, there are several levels of mockup implementation. Each has its place, depending upon the needs of the designer and perhaps the time and money available to create the mockup. The following descriptions and use suggestions provide a guide to help designers decide upon which level of mockup implementation they need to accomplish their objectives.

1. Paper Mockup

The paper mockup is the least expensive and time-consuming to prepare. It is constructed by cutting out scaled, two-dimensional elements and/or a drawing and placing these on a plan view of a work area or on a wall representing the vertical surface of some cabinet that will eventually hold the proposed control panel.

For a miniature-scale workplace or facilities arrangement problem, one can create an excellent first-step evaluation of various alternatives; i.e., one can try out several possible arrangements of consoles, equipment racks, desks, or other furnishings to determine the arrangement that provides for the fewest operator steps, the best clearances for traffic flow, the best viewing angles, and so on. Such mockups can be made out of either paper or cardboard (the latter tends to lay flat more readily and is less affected by drafts, which often blow paper cutouts off the table). Typically, one starts out with a substantial baseboard marked off in square feet (to whatever scale is chosen by the designer). Then the equipment items are cut out of either paper or cardboard (to scale) and are labeled. The cutouts are then manipulated into various alternative positions until the best arrangement is obtained, considering all the critical factors of traffic flow, intercommunication between operator

stations, access for maintenance, and a general feeling of spaciousness or lack of clutter. This type of mockup is very effective for architectural planning, such as for determining room size or equipment arrangement or even for revamping office space or work space in an existing building or office.

The paper mockup is also suitable for initial examination of full-scale relationships. For example, one can place a full-scale drawing of a proposed control panel on a wall in order to determine at what height the panel should be placed or what the best orientation of the panel would be from a nominal operator position (either standing or sitting). Such a mockup can also be used for initial evaluation of the arrangement of controls and displays; i.e., one can simulate going through the proposed sequence of panel operation and quickly discover that items are arranged in the drawing in such a way that too many hand crossovers occur, that things used most often are too far out of reach, that one tends to cover a display while operating a control, or that the proposed labels and the proposed arrangement of panel elements are confusing, making the panel elements hard to locate quickly. Since no construction is required, it is easy and inexpensive to try several alternatives until the best one is found. The paper wall mockup is also useful in establishing the size of labeling required for viewing from alternate positions, such as from across the room. Finally, this inexpensive approach provides the first opportunity to evaluate space and reach problems using test subjects at the extremes of body size to make sure that both large and small operators are accommodated equally well.

It should be emphasized that the paper mockup is not to be regarded as a plaything that is beneath the dignity of the practicing professional. Experience has demonstrated many times that more costly mockups have not provided any more useful information than a paper mockup would have provided, and considerable time is often wasted on more exotic mockups, not to mention the cost. Finally, it should also be noted that a paper mockup will provide important criteria for developing more sophisticated mockups later on during the design program.

2. Soft Three-Dimensional Mockup

A soft mockup is one that is made of wood and cardboard or of special inexpensive laminates such as Foam Core. The term "soft" relates primarily to the fact that the materials are easy to cut and assemble, as opposed to metal. The general approach consists of creating a simple wood understructure, to which sheets of cardboard or Foam Core are attached to provide the "skin," e.g., a console, panel rack, or workplace enclosure. To these skins are attached panel drawings or simulated controls and displays that may be cut out of stiff cardboard upon which pictorials of instruments or controls and/or function labels are mounted (to give an added three-dimensional appearance not unlike that of the real hardware elements).

The principal value of the full-scale three-dimensional mockup is its ability to represent spatial interrelationships between an operator and the controls and displays he or she must monitor and manipulate, arm reach and viewing angle parameters, and clearances for feet,

legs, and body (including seats). For early stages of design analysis, it usually is not necessary to enclose a console completely, for example. Only the front of the console is important. The back of the console need not be enclosed, and thus it is not necessary to spend the time and money required to mock up several alternative console shapes.

Three-dimensional miniature-scale mockups can be made of wood, preferably some lightweight wood that is easy to fabricate (e.g., balsam). Mockup elements (consoles, panel racks, desks, chairs, machine tools, etc.) are more easily manipulated if small magnets are embedded into the base of each piece and if the baseboard representing a floor space or area is made of a suitable board upon which a metal plate is mounted. In this way, the equipment and furnishing models are easily moved about and yet stay in place.

A second-level soft mockup may be desirable once the general pattern of a console has been established. That is, in place of the paper or cardboard display and control simulations, actual instruments can be mounted on the control panels. These displays and controls can in turn be made active to various levels, depending on the purpose of the simulation. That is, instruments can be wired to demonstrate illumination, or they can be instrumented to the extent of tying into a computer program, which allows one to simulate and evaluate alternative display and control system parameter concepts.

A third level of sophistication for the soft mockup consists in mounting the mockup on dynamic platforms that simulate, for example, the motions of an aircraft, space vehicle, or ship. Generally, one would not go to this level of time and cost unless there was some solid indication that the dynamic aspects of the operator-control situation were especially critical to the ultimate functioning of the human-machine system. It becomes obvious that this level of mockup begins to approach the capability of becoming a training simulator. Therefore, one can equate the time and cost in terms of both a design tool and an eventual training tool that may be required by the customer.

Although soft mockups are recognized as a useful tool for examining three-dimensional operator-station problems, it should be noted that they are also extremely useful for examining the problems of maintenance technicians. Once again, both miniature-scale and full-scale mockups have their place, for the reasons noted above. For example, the miniature-scale model provides an excellent method for examining proposed layouts for production and for service and maintenance facilities. Using the same approach discussed above, scale models of prime vehicles, support vehicles, structural features, etc., all can be easily viewed on the small-scale simulation. An added benefit of the scale model is that, when different groups of people are involved (such as the contractor and the customer, who may be many miles away), photographs can be taken of several alternative layouts that have been tried, and these can be mailed back and forth to increase the level of communication between the interested parties. In fact, the photographic technique provides an excellent method for preparing for a design review briefing. For example, the designer can take pictures of each alternative arrangement that he or she has tried and evaluated. By making these into slides, it is possible within a few minutes to run through and explain the rationale for making a final arrangement recommendation. This saves a lot of questions from those who would like to know whether the designer has explored all possibilities.

Finally, in the case of mockups that will not be involved in any kind of dynamic evaluations (i.e., mockups which are static, which will not have to support heavy components, or which will not be sat upon or leaned against by persons involved in the evaluations), it is important to use materials that are easily and quickly modifiable by the designer so that it is not necessary to wait for a mockup specialist to make changes. On the other hand, some mockups will have to have sufficient structural integrity to support heavy components, persons who may climb upon them, or accelerative forces from moving platform dynamics. These generally are all-wood structures and in certain areas are covered with a plywood skin.

3. Hard Mockup
Generally speaking, the so-called hard mockup is used to define the detailed assembly aspects of a final production design. For example, the hard mockup will be made of metal and will include the precise details of the inner structure and outer skin, which the designer, along with the production engineer, will explore and for which they will define types of attachment hardware for internal components, external components, trim, etc. A good example is the fuselage mockup that is used to define the routing and attachment of cables, hydraulic lines, pneumatic lines, control cables, etc. Although the designer generally prescribes where these components go and in general how they will be attached and connected, it is not certain what obstacles there may be within the basic fuselage structure. Thus the hard mockup provides a precisely scaled breadboard, so to speak, on which the designer can determine how best to route and fasten components. Once these decisions are made and final routings and fastenings are actually mounted in the mockup, the mockup becomes a reference for production-line personnel. Typically these hard mockups are retained for the life of a production run and sometimes beyond (to be used for later modification analyses).

The hard mockup also provides one other important contribution. In deciding how to assemble large, complex elements of a hardware system, it becomes important to work out a plan for assembly that minimizes conflicts among assembly operations. For example, in very tight spaces, various assembly operations cannot all work in the space simultaneously, and the mockup allows one to determine the most efficient order in which to assemble components.

From a purely human engineering point of view, the hard mockup provides an excellent opportunity to evaluate specific human-machine interface problems. These include maintenance tasks (accessibility to components and parts that have to be inspected, removed and replaced, calibrated, or adjusted in the field).

4. Other Mockups
STYLING BUCKS AND MODELS The vehicular industry has long used and depended on the clay model to "shape" or "style" the exterior of a new-model car, truck, or bus. Over a basic frame, modeling clay is laid and shaped to represent the proposed exterior lines of the new model (and alternatives). Such models are made in both miniature- and full-scale versions. A special clay material allows the model to be painted so that the full scope of appearance elements can be evaluated for eye appeal.

For smaller products, either the structure and clay approach can be used, or package appearance design can be modeled from other more substantial materials such as fiberglass or thermo-setting plastics. This type of model provides sufficient structural integrity to allow more specific user testing; i.e., a model of a hand tool, hair dryer, or electric knife can be picked up and manipulated to test manipulability and ease of handling.

PORTABILITY MODELS Many small products require handling by the user, operator, or maintenance technician; i.e., they must be picked up, positioned, operated, donned or doffed, transported, and set down again. Before final package decisions are made, such models provide an important step in validating concepts before the product is finally produced. Such models should simulate the dimensional and dynamic aspects of the product so that one can determine whether the expected range of users can, in fact, manipulate the package with ease and surety. Often, until the package configuration is tested in full-scale model form, one cannot be sure whether the weight and balance characteristics of the proposed package will stress users or cause them to drop the package or lose their balance.

ACCESSIBILITY MOCKUPS Although many military projects require key demonstrations of accessibility for maintenance at certain points during system development, these tend to pertain to major access points in an aircraft fuselage, a tank shell, or other prime hardware unit. These represent only the minimal areas where maintenance access mockups should be considered. For example, any small piece of equipment such as a radio, radar, or sonar unit or a piece of test equipment is a subject for verifying maintenance accessibility by means of a mockup. One should consider early mockups before packaging constraints preclude optimizing access; i.e., by means of an early mockup of various package configuration alternatives, one might be encouraged to change the shape or orientation of the package in order to provide a more convenient opening position when the unit is finally installed within close-fitting quarters. A second level of mockup could examine the positioning order of components so that a technician will not have to remove several interfering components to get at the one that is most often expected to need inspection, removal, and replacement. These mockups are especially useful for evaluating various types of slides, fasteners, latches, and handles, as well as for various approaches to the articulation (for accessibility) and/or the removal and replacement of chassis and printed circuit boards.

How to Determine What Type and Level of Mockup to Use

There are no hard-and-fast rules for deciding what type of mockup to use or to what level of sophistication one should go in developing the mockup and/or operational simulation. The following general guidelines may be helpful.

AVAILABLE TIME Although one should make every effort to find the time to "be sure" (i.e., by mocking up and verifying a complex, three-dimensional concept before committing it to final design), the time available before a design has to be completed and the product delivered may make it impossible to develop and evaluate the mockup. There have been instances where an extensive mockup and design simulation program was completed months after a system had been delivered to the customer. In order to avoid such errors, it is vital that both the customer and the contractor recognize and prepare reasonable schedules so that, if a mockup or simulator is vital, its capabilities can be used in time to influence hardware design.

CRITICALITY OF THE HUMAN-MACHINE INTERACTIONS The following are considered critical interactions:

1. The operator and maintainer working space is limited because of external confining restraints (e.g., a small cockpit, close maintenance quarters, low overhead, or limited crawl space). Such space limitations are particularly critical if the worker or operator may have to wear heavy protective garments. The space should be evaluated with large test subjects wearing the most constraining type of garment.
2. Escape envelope constraints are especially critical for obvious safety reasons. Mockups should be used to evaluate not only the envelope shape and size but also the position of the exit with respect to potentially difficult conditions under which the escape envelope must be traversed and the position of the final exit with respect to opening and departure (e.g., an exit on the top of an aircraft fuselage may be too high for a small person to get to, especially if high g forces are working against the escapee). Escape system and procedure mockups are vital for determining whether escape-time criteria can be met by a particular design configuration.
3. Visibility constraints and/or adequacy should be verified using three-dimensional mockups. Although general anthropometric criteria provide ball-park estimates of how to arrange operators and equipment, windows, glare shields, etc., the final verification requires *in situ* examination by selected ranges of subjects of both large and small stature. Sight lines to both internal displays and external viewing requirements need verification before a final operator-station configuration is accepted.
4. Reach constraints should be verified for certain operator and maintainer workplaces, especially if the operator or maintenance technician has to be restrained

by a safety belt or limited access space. The question includes the reach limits not only of small individuals (possibly constrained by special garments) but also of large individuals, who may not have sufficient clearance to retract an arm or leg.

5. Lighting parameters are often critical where it is very important to provide low but even illumination of all visual displays, labels, and controls or where it is possible that reflections and glare sources may interfere with proper display or external, out-the-window viewing. Such mockups should be evaluated under all the expected ambient lighting conditions. It is wishful thinking to believe that one can identify and adjust a design for all possible glare or reflection problems.

6. Ingress and egress mockups are an important consideration, not only for the questions of emergency escape noted above, but also for normal entry and exit. This is especially important in the case of vehicles for which older and disabled individuals are expected to be among the user population.

7. Control and display arrangement for ease of location, identification, and speed of use should be mocked up and evaluated in real time by a number of typical user subjects. This is especially important when the control operation is complex and when mistakes may be costly to the final operation. Although a cursory link analysis of a preliminary layout drawing provides the first indication of whether a panel layout appears to be logical and convenient, it may not tell the full story; i.e., when the panel is associated with other displays and controls and possibly a constraining seat, the otherwise adequate panel arrangement may prove to be less effective than it appeared in the drawing analysis.

8. Component-rack interactions may present difficult mobility problems for the maintenance technician, and therefore they should be examined in the three-dimensional mockup setup. Typical problems that show up in the mockup evaluation include the following: the technician cannot bend over far enough to lift a chassis out of the lower portion of a rack; there is not enough clearance to open an access door fully; an access opening is so small that the technician's hand or arm blocks the view of the object he or she is trying to reach or manipulate; and there is an intervening hazard when the technician reaches across a moving belt or gear to adjust or remove some component.

UTILITY OF A PARTICULAR MOCKUP LEVEL TO PROVIDE NECESSARY ANSWERS There are those who swear by mockups, just as there are those who think that mockups are a waste of time and money. Mockups are often made more elaborate than necessary, usually because an individual manager or engineer likes to have the most lifelike representation possible to show a customer. Although this is impressive and may have value from a marketing point of view, it may be a waste in terms of its design verification value. On the other hand, making the mockup too minimal may prevent one from finding answers to the questions outlined above. The designer must decide what kind of mockup will provide the needed answers; any expansion beyond this requirement is probably a waste of effort and money.

This problem is particularly difficult to resolve in the case of mockups that tend to be real-life simulations. That is, how far to go with simulation is always a moot question. In general, one can never completely simulate the real world under any circumstances, short of creating the actual article and testing it in a real-world environment. Basically, that is the role of a prototype article, e.g., a prototype aircraft, a prototype automobile, or a prototype piece of test equipment. Short of this, mockups and simulations of partial task elements are often more cost-effective.

MOCKUP ADJUSTABILITY To be maximally effective as a design evaluation tool, a mockup should be made as adjustable as possible. As noted above, there often is a tendency to feel that a mockup can be made to reflect all the "right" features the first time and that the only purpose is to prove that one's design analysis is correct. Almost without fail, one finds that changes are required after the evaluation. However, unless the mockup is constructed in a fashion that allows for easy modification, it may be necessary to fabricate an entirely new one. Before constructing any mockup, therefore, look ahead and identify all the features that you may suspect could require adjustment during or after the first evaluations. This will allow you to make these changes and reevaluate them with a minimum of time and cost. It will also allow you to demonstrate quickly to customers why you have chosen the final configuration you have— merely by showing them the several versions you tested.

Miniature-Scale Mockups and Models

Miniature-scale mockups and models are especially useful for evaluating the multicomponent layout and arrangement of large elements such as clusters of buildings and large structures, weapons and troops in a battle configuration, multiple aircraft and support equipment on a flight deck, and equipment, machines, and furnishings within a building, office, laboratory, or factory. Depending on the overall number of elements, one should select the scale to fit the evaluation purposes. Models need not be of fine detail, but they should be fairly precisely scaled so that traffic flow and work-space clearance can be realistically assessed. It is suggested that ⅛-scale models be used for interior room arrangements if it becomes desirable to evaluate any detail on model equipment or furnishings (position of displays, lines of sight, illumination possibilities, etc.).

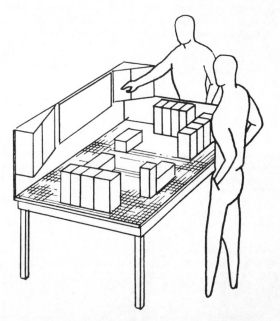

to provide vantage point to observe total arrangement proposals

As long as the mockup representation involves a flat floor, it is suggested that a metal plate inscribed in a 1-ft (0.3-m) grid be used as the primary reference base. This makes it easy to assess clearances quickly.

Model equipment, furniture, and machinery should be magnetized so that, although these elements stay put and are not easily disarranged or knocked over, they are still easily moved from one place to another.

Use miniature-scale models to:

Evaluate traffic flow
Establish visual sight lines
Route utilities
Evaluate workplace convenience and maneuvering space
Evaluate production-line flow
Position common-access elements
Make a preliminary lighting and color analysis
Determine the best escape routes
Determine access for maintenance and service
Decide on preliminary decor

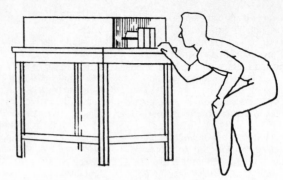

for preliminary examination as actual operator would see it

Use of Drawings and Paste-Ups

Early evaluation of proposed control panel positioning and layouts can be accomplished inexpensively by pinning a drawing on a wall or, better yet, by pinning or attaching self-adhering panel element patterns on a cork or felt board. If a metal wall is available, similar results can be obtained using two-sided tape or flexible magnetized tape on the back of panel elements. It is recommended that independent panel elements be placed over the panel drawing because this permits moving the elements around on the board.

As illustrated by the accompanying sketches, both standing- and sitting-operator relationships can be evaluated. The following key issues can be resolved by these techniques:

Identifying reach problems
Analyzing operational sequence and functional grouping
Determining proper spacing of components and positioning of proposed labels
Establishing the best height for visual displays and determining whether a hand or arm obscures a display

When it appears desirable to change the plane of a particular panel, an inexpensive independent structure can be devised upon which the drawings and/or cutouts can be pasted (as illustrated in the accompanying sketch). A wooden structure can be covered by cardboard or Foam Core to provide sufficient panel rigidity. A light metal panel can be easily attached to the structure and/or Foam Core surfaces, making it possible to utilize the magnet technique for attaching panel components independently.

Adjustable, Erector Set Mockups for Operator Control and Display Positioning Definition

As opposed to establishing control and display positioning using general reach criteria, it is highly desirable to establish optimum positions by placing an operator in a three-dimensional framework, unobstructed by preconceived consoles, racks, bulkheads, etc. As the accompanying illustrations show, an open framework provides an ideal method for adapting controls or displays to the operator. Control and display elements should be mounted on fully adjustable supports so that the control or display panel can be moved around until the best position is found in terms of line of sight or reach.

Controls should be mounted in such a way that the control device can be moved through its intended excursion, especially joy sticks, steering wheels, foot pedals, etc.

It is also possible to attach key visibility elements (window posts or frames, rearview mirrors, etc.) to evaluate fields of view.

A cockpit or farm machine mockup

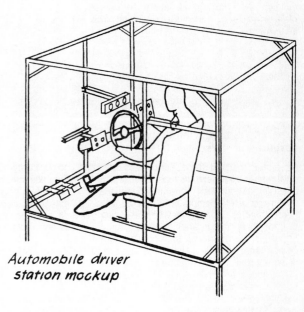

Automobile driver station mockup

As the accompanying sketches illustrate, service and maintenance questions can also be evaluated with the Erector Set mockup approach.

Service equipment rack mockup

Portable package/handle-locating mockup (note sandbag weights)

Simple Graphic Art Techniques for Mockup Fabrication

One should seek the simplest techniques commensurate with demonstrating the significant human-machine interface characteristics that are to be evaluated. Two of these are illustrated by the accompanying sketches.

Simulated three-dimensional visual control components made from readily available materials, if executed properly, give a realistic impression of the components that an operator would expect to see on a control panel. The clear plastic over the dial face allows one to anticipate annoying reflections.

The use of movable components makes it possible to try more than one arrangement.

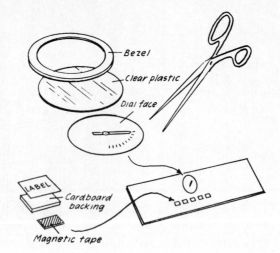

An inexpensive yet realistic-looking console can be constructed of a wood frame and smooth skin; the latter can be painted (using a water-base paint) to reflect the desired color scheme and demonstrate the desired contrast between the sheet metal and the control panels (assuming they are of a different color or shade).

Panels should be taped both inside and outside to provide structural integrity. When properly taped, the mockup will support considerable weight (e.g., actual metal panels and small instruments).

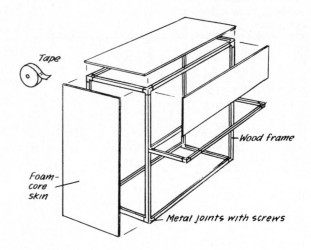

Experimenting with Alternative Panel Layouts

More often than not, the designer has to utilize off-the-shelf hardware for most control panel designs. It is important, therefore, to recognize that many of the components may be larger "behind the panel" than they appear to be on the front side of the panel. In making cutouts of standard instruments, switches, and other components, it is wise to draw in (using dotted lines) the behind-the-panel clearance requirements so that one is not apt to arrange components closer together than they can actually be placed when it comes time to mount the actual instruments.

In experimenting with alternative panel layouts, one should make a checklist of the principal human engineering features that should be kept in mind, i.e., functional organization, sequence of use, frequency of use, and primacy or importance. In addition, however, one can also examine a number of factors that are

not in the typical human engineering guides. For example, although there are criteria for spacing to prevent inadvertent activation and criteria for size of label letters, etc., spacing often is a matter of general appearance, e.g., balance, symmetry, and absolute clarity. One can "see" these characteristics only when experimenting with several different arrangements and spacings. For example, it may become obvious that, because the panel has to be very small, crowding obviously creates confusion. By making slight alterations and/or adding separator lines around certain related functions, one can alleviate the confusion. One can also tell rather quickly when the size variation among labels is not sufficient to provide an immediately clear indication of function levels. Even though criteria have been established by human engineers, these sometimes need to be adjusted.

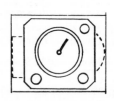

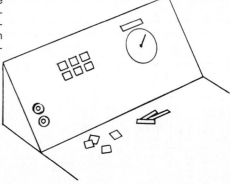

Human-Machine System and Environment Simulation as a Design Tool

Any physical mockup is in a sense a simulation of the human-machine environment. As the system becomes more complex, however, it is often desirable to enlist the aid of more exotic apparatus, coupled with computer-aided scenario generation. As the accompanying sketch illustrates, one may require simulation (and control) of both external and internal environmental parameters, since all these influence how well the human-machine system functions in various modes. Not shown is the possibility of including dynamic properties such as oscillation and movement of the subject oper-

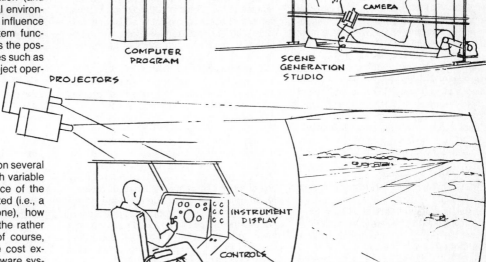

ator. The extent of realism depends on several things, including how important each variable may be to the eventual performance of the operator, how well it *can* be simulated (i.e., a poor simulation is worse than none), how much time is available to develop the rather complex simulation system, and, of course, whether the cost is justified. If the cost exceeds the cost of the ultimate hardware system, it usually is not justified. Avoid the temptation to create an exotic simulation just because it is a design challenge.

Visual Display Element Evaluation

Typically, one should rely on the criteria provided elsewhere in this handbook to determine the type of characters to use for labels. However, display nomenclature may be difficult to establish without evaluating the aspect of understandability. A simple tachistoscopic slide presentation is suggested, wherein one can test several alternative nomenclatures, abbreviations, or pictorial symbols to determine which ones lead to the fewest errors in interpretation. As illustrated, several subjects can be used simultaneously to evaluate the alternative display nomenclature, symbols, or even other visual parameters, such as size and color.

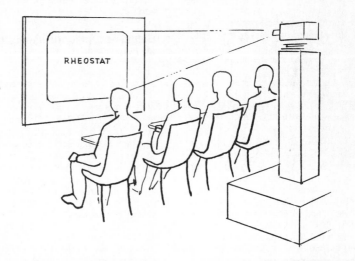

Choosing Subjects for Mockup Evaluations

Although there may be good reasons (i.e., time and funding constraints) for limiting mockup evaluation to the designer and his or her immediate associates, the results may be far from satisfactory. There are several reasons for this:

1. Designers may be biased toward a design they have just created; i.e., because it is their own "brainchild," they really do not want to see any problems that the design may create for the user.

2. Designers are not always representative of the user population in terms of either physical or psychological characteristics; i.e., a designer may be extremely large or small, very strong or very weak, or particularly skilled in terms of equipment manipulation and understanding, or the designer may tend to "read into" the operation what the design is intended to demand because he or she already knows what is expected of the user.

3. A male designer may not understand how a woman will react to a design.

4. Designers, because they are adults, may fail to recognize or understand how a child or younger person will react to a design.

5. Designers typically assume that all consumers will react to a product design in the same way they do (after all, they are also consumers). However, since a design may eventually be used by people with different ethnic, cultural, or language backgrounds, things about the design that appear perfectly clear and/or desirable to the designer may be completely foreign and undesirable to many people who eventually become users of the product.

Guidelines for Selecting Mockup Evaluators

The following guidelines are provided to aid the designer in selecting subjects to use as evaluators.

POPULATION Select evaluators directly from the expected user population, such as military personnel (as opposed to civilians), adults (as opposed to children), or members of a particular professional or vocational group.

BODY SIZE Select evaluators from specific or general populations on the basis of recognized anthropometric surveys. For example, several professions (Air Force, Army, and Navy personnel; astronauts; bus drivers; etc.) may be limited by anthropometric selection criteria; i.e., they will be only so tall or so heavy because a particular agency has established selection criteria that preclude taking applicants outside these dimensional ranges. On the other hand, not only do civilian populations include persons with a wider range of body sizes, but also the range is expanded because of inclusions of both males and females and both young and old people.

STRENGTH When the mockup evaluation involves lifting and carrying the product and/or manipulating controls that may have fairly heavy force requirements, the evaluator population should represent mainly the weaker members of the expected user population.

INTELLIGENCE, EXPERIENCE, AND TRAINING In certain cases, it would be wasteful of human capability to try to make a product acceptable to everyone; i.e., some people would never be able to cope with very complex equipment operations, even with extensive training. When the operation of the proposed product normally would be restricted to specialists already known to have the requisite knowledge and skills, there is no need to compromise the proficiency of the human-machine system (i.e., to make it unduly simple). In such cases, evaluators should be chosen who are already knowledgeable about the proposed design and therefore can evaluate the mockup or simulation from a skilled operator's point of view.

On the other hand, if the product may be operated by unskilled individuals (and therefore may create a critical training requirement), evaluators should be chosen from an unskilled group in order to determine what there may be about the design that is confusing or difficult to manipulate. This provides an opportunity to determine what modifications may be necessary to make the design more "trainable" and less apt to create special personnel selection requirements for the eventual customer.

PRODUCT ACCEPTABILITY In evaluating a proposed design (via a mockup), one should consider a highly varied sample of evaluators; i.e., the sample should include people of both sexes and of all sizes, ages, and backgrounds in order to allow for the random inclusion of personal attitudes and opinions, physical difficulties, etc. Although such subject samples should be generally random in terms of all these varying characteristics, it is often necessary to make sure that specific human characteristics are represented. This is important whenever the designer suspects that certain characteristics of the design may create special problems for certain members of the user

population. For example, in evaluating seat belts, it is known that large and small people have certain kinds of problems. Although one wishes to obtain a general evaluation of how people may expect to use a system or how they may react to certain features of convenience, it is important to make sure that a sufficient number of the large and small evaluators evaluate those features of any belt system that cause belt systems not to fit properly. Thus, although the sample should generally represent the random attitudes of a broad sample of users, one should add a few subjects who are representative of the extreme anthropometric characteristics of the general sample.

TECHNOLOGICAL VARIABILITY There is a tendency to believe that just because we are a technologically oriented society, all product users understand and can cope with modern machines. Studies have shown that this is not true. In some societies, people have not had the opportunity to grow up with modern technology; therefore, not only are they unable to cope with the demands of modern machines, but they also may completely reject machine-age products. This presents a difficult problem for designers, for seldom do they have an opportunity to study the population in question, nor are there typically any study data to help them determine what a particular population will do or expect. The principal recourse under these circumstances is to search for subjects who may be recently from the area in question, using these evaluators as a screening group to determine whether there will be major difficulties as a result of certain characteristics of the design that have been mocked up for evaluation.

Guidelines for Determining Number of Mockup Evaluators

First of all, the designer may be confronted by the problem of convincing management that it is important to evaluate a mockup in more than a cursory sense. For example, many management people believe that *they are the final judge* in determining whether a design is satisfactory. Although it is probably true that some manager will have the final say, it is dangerous to assume that such a person has the knowledge to make judgments regarding effective human-machine performance. Convincing such a manager to rely on a more scientific approach to product evaluation, one that reliably reflects the specific user's response to the design, requires considerable diplomacy on the part of the designer. Consider the following in deciding whom to use as evaluators and how many to use.

EVALUATING REACH AND CLEARANCE Although a fairly large, random sample of subjects is required when one wishes to establish some statistical distribution for reach and clearance, in most cases it is satisfactory to evaluate these parameters in a proposed design (via a mockup) using a minimum of about five large subjects to evaluate clearance and about five small subjects to evaluate reach. One should select these groups on the basis of the particular population for which the product is being designed, such as a military or civilian group. For clearance, select five subjects whose stature is at approximately the

95th percentile and whose weight is also at about the 95th percentile. For reach, select five subjects whose stature is at about the 5th percentile (for a mixed user population, use 5th-percentile females). For all practical purposes, selecting five subjects at each end of the stature range will provide sufficient variability in the other characteristics to provide an adequate mix of characteristics that also relate to clearance and reach, e.g., variations in hip breadth, buttock-to-knee length, knee height (sitting), head height (sitting), eye height (sitting), head height, shoulder height, shoulder breadth, abdominal depth, and lengths of upper and lower arm segments. In addition, one usually can depend on a reasonable variation among 5 or 10 subjects in terms of overall body and limb mobility.

EVALUATING WEIGHT CHARACTERISTICS OF A PORTABLE PACKAGE The minimum number of subjects required to evaluate the weight characteristics of a package mockup is about 10. One should select relatively small subjects since not only is the size of the subject a general indicator of overall strength, but also the ability to lift is a function of the length of a person's arms, the size of his or her hands, and the general girth of the individual's abdomen, chest, and shoulders. One or two of the subjects should be fairly fat, since the fat individual may have difficulty holding the package close to his or her chest and stomach.

EVALUATING CONTROL MANIPULABILITY Since a number of significant variables influence how well an individual may be able to manipulate a control device, a minimum of 20 subjects should be used. In general, the subject sample should be selected at random, since one hopes to sample a variety of individual differences, including anthropometric differences and differences in mobility, dexterity, steadiness, strength, sensorimotor proficiencies, and psychological expectations. Depending on the expected user population, both age and sex should be represented about equally throughout the sample; i.e., the subject sample should consist of half males and half females, and there should be a wide distribution of ages.

TEAM PROCEDURE: MOCKUP AND SIMULATION EVALUATIONS When more complex mockups are used to simulate team function and procedural response, a minimum of three complete teams should be used; i.e., as opposed to having an evaluation made by a single team of subjects, repeat the evaluation with at least two additional teams of subjects. This is desirable because one is never sure whether the first team is either unusually good or unusually poor, nor can one be sure, if the performances of the two teams vary significantly, which one's performance is truly representative of the system's quality in terms of design effectiveness. The response of the third team provides a means for breaking a tie, in addition to indicating which response performance is probably more representative of the design quality.

COMFORT AND CONVENIENCE EVALUATION An evaluation of the comfort and convenience aspects of a product or system mockup is perhaps the most difficult type of evaluation to perform, because in spite of all attempts to focus the evaluator's attention on significant design areas, each individual makes a different internal interpretation of what constitutes comfort or convenience. Therefore, it is recommended that a minimum of 50 subjects be used for this type of mockup evaluation. It is further suggested that, in performing this kind of evaluation, a special evaluative approach be taken, i.e., one in which subjects are forced to focus on, and respond to, specific, predetermined aspects of the design, one at a time—or at least short groups of aspects that can be separated in terms of operational steps. An example would be the case of evaluating seat belts in an automobile. Here, one might segregate the several typical use sequences, i.e., entering the vehicle and donning the belt, adjusting the seat and closing the door, simulating operating the vehicle (looking out the windows, reaching for the controls, looking at the displays, etc.), and finally doffing the belt and getting out of the car. Upon completing each of these sequences, the subject would be asked to rate various features of the belt system as they pertained only to that particular operating sequence.

Also important in this area of comfort and convenience are dynamic and environmental factors. Although one can obtain a general, first approximation of consumer response in the static mockup, it is probable that a person's responses would be modified after actually experiencing the seat belt under more realistic conditions, i.e., after riding for awhile over various road conditions.

Establishing the Basis for Test Subject Selection

PRODUCT FORM AND FIT PROBLEMS
When the problem involves physical relationships between an operator and some hardware element (seat, console, cockpit, furniture, etc.), test subjects should be selected to represent the critical dimensional limits of expected operators. First, it is important to have several subjects who represent the largest people who may have problems of clearance and several subjects who represent the smallest people who may have problems of reach. Second, there should be a scattering of subject sizes in between the largest and the smallest in order to provide assurance that some in-between dimensional incompatibility is not overlooked. In a sense, we have a stratified sample of sizes because we know that certain physical accommodation problems are more critical at the extremes.

It is standard practice in the military services to attempt to fit 90 percent of the population, i.e., from the 5th- to the 95th-percentile population of a particular military sample. This means, however, that, in the event the product is to be used by both male and female operators, the test sample should range from the 5th-percentile female through the 95th-percentile male (as opposed to ranging from the 5th- to the 95th-percentile male when only male operators are expected).

OTHER PRODUCT FACTORS
When the questions to be answered do not relate to the physical size of the expected user population, a more random sample of test subjects should be used. For example, when we are investigating sensorimotor response relative to some proposed display and control system, body size has less to do with the subject's response, and therefore we need a random sample of subject skills, experience, motivation, and attitudes in order to assess the average response of an expected user population.

If the problem has something to do with human strength, it may be desirable to select more subjects who are weaker, since these are the users who may have more difficulty with problems of control force, lifting, and so on. It should be noted that one cannot determine how much strength a subject can apply merely on the basis of his or her size, although it can generally be assumed that females are weaker than males.

If the product may be used by older people, it is important to include a number of elderly persons, both male and female, to reflect the compatibilities or incompatibilities of the proposed product design with typical age-related infirmities, e.g., poor vision, reduced mobility and strength, and slower response time.

Similarly, if handicapped users are expected, a variety of handicapped test subjects should be included to reflect the appropriate range of infirmities and difficulties they may have, such as being confined to a wheelchair.

Ethnic differences, such as differences in language or technological background, may be an important aspect of human factors testing, and the appropriate selection of test subjects who represent unique design interface barriers must be considered. For example, a test to determine how intelligible certain pictorial symbols are to people of various nationalities might be appropriate.

GENERAL PRODUCT ACCEPTANCE
For most consumer acceptance tests or surveys, it is desirable to take a completely random sampling approach since the normal sales objective is to reach as broad a population of purchasers as possible. In such cases, it is up to the customer to establish the limits to which he or she expects to promote the product, i.e., to the world, to a specific country, or to a specific part of a country. It should be observed that this type of sampling generally relates to consumer opinions and attitudes, which change frequently. Therefore, not only should sampling be approached on an area-by-area basis, but also a successive series of surveys should be carried out to allow for opinion shifts. A good example is provided by the automobile industry. Some people have to get used to a new model of an automobile; i.e., the new body style may be sufficiently radical to "turn them off." However, after 6 months to a year, they are accustomed to the new style, and it becomes the reference by which they measure satisfaction with other new models.

Human Physical Dimensions for Design of Equipment and Furnishings

PHYSICAL DIMENSIONS

The science of measuring the human body is *anthropometry.* Anthropometrists have been measuring a wide variety of people in many ways for many years. More than 350 different measurements have been taken at one time or another. Although measuring techniques have become more or less standardized, there is no assurance that each anthropometrist measured his or her particular sample of subjects, or each particular dimension, in the same manner. Thus one has to assume a certain caution in comparing data from various populations.

A great share of anthropometric data has been generated for the purpose of scientific comparison of populations, and not necessarily for the purpose of providing design-related information. This is evident from the fact that subjects are measured in stiff and unnatural positions, and in the nude. Also, since anthropometric surveys are extremely expensive, they are not repeated periodically to update information as people and populations change. Because of these and other unique conditions, it is important that the user of anthropometric data recognize the limitations of such data and treat the information merely as a point of departure. That is, anthropometric data are most useful as an aid in selecting a sample of "live" test subjects, who then should be used to test any proposed design.

Unfortunately, not all surveys include all the dimensions one might need for a given design problem. The most complete surveys have almost always been conducted on military personnel, and most of these have dealt with males.

In spite of these shortcomings, designers should make a point of referring to anthropometric data whenever a design involves a "fit" problem. Also, care should be taken to refer to data that are representative of the pertinent user population, since a particular survey may not have examined the dimension in question and thus may provide no data on it. To assist the designer, the following information on body dimensions has been organized according to the population groups in which product designers are most commonly interested. It will be noted that there are some obvious gaps in the tables where data for a given dimension and population are unavailable. Also, it should be pointed out that, in some cases, a substitute value has been entered when (in the author's opinion) a similar population seemed to justify using a particular dimensional value for the population in question. Although such liberty is often frowned upon by conservative anthropometrists, it is excused here on the basis that *none of these data should be used directly for design purposes, but rather as starting points to establish requirements for selecting a dynamic (live) test subject sample.* There is no substitute for a good, full-scale mockup evaluation to validate "fit" of the human body.

HUMAN PHYSICAL DIMENSIONS
Anthropometric Reference

**BASIC U.S. ADULT CONSUMER
ANTHROPOMETRIC REFERENCE***

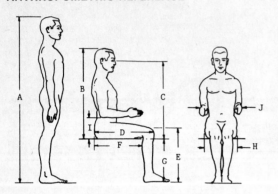

A – Standing Height
B – Sitting Height
C – Seated Eye Height
D – Upper Leg Length
E – Knee Height
F – Seat Length
G – Seat Height
H – Seat Width
I – Elbow Rest Height
J – Elbow Room

Basic Percentile Dimensions
of U. S. Adult Population
(18 to 79 years)

%	A	B	C	D	E	F	G	H	I	J	Weight
5	63.6 59.0	33.2 30.9	28.7 27.4	21.3 20.4	19.3 17.9	17.3 17.0	15.5 14.0	12.2 12.3	7.4 7.1	13.7 12.3	126 104
10	64.5 59.8	33.8 31.4	29.3 27.8	21.8 20.9	20.0 18.2	17.9 17.3	16.0 14.2	12.5 12.7	8.0 7.6	14.3 12.9	134 111
20	66.0 61.1	34.4 32.2	30.0 28.4	22.3 21.3	20.4 18.6	18.4 17.9	16.4 14.7	13.1 13.3	8.5 8.2	15.0 13.5	144 118
30	66.8 61.8	34.9 32.6	30.5 28.7	22.7 21.7	20.7 19.1	18.8 18.2	16.7 15.1	13.4 13.6	8.9 8.5	15.5 14.1	152 125
40	67.6 62.4	35.3 33.1	30.9 29.0	23.0 22.1	21.1 19.3	19.2 18.6	17.0 15.4	13.7 14.0	9.2 8.9	16.0 14.6	159 131
50	68.3 62.9	35.7 33.4	31.3 29.3	23.3 22.4	21.4 19.6	19.5 18.9	17.3 15.7	14.0 14.3	9.5 9.2	16.5 15.1	166 137
60	68.8 63.7	36.0 33.8	31.7 29.6	23.6 22.6	21.7 19.8	19.8 19.2	17.6 16.0	14.3 14.7	9.8 9.5	17.0 15.6	173 144
70	69.7 64.4	36.5 34.2	32.0 29.8	23.9 22.9	22.0 20.1	20.1 19.5	17.8 16.3	14.6 15.1	10.2 9.7	17.5 16.3	181 152
80	70.6 65.1	36.9 34.6	32.5 30.2	24.4 23.4	22.4 20.5	20.5 19.9	18.2 16.6	14.9 15.6	10.6 10.1	18.1 17.1	190 164
90	71.8 66.4	37.6 35.2	33.0 30.7	24.8 24.0	22.9 21.0	21.0 20.6	18.8 17.0	15.5 16.4	11.0 10.7	19.0 18.3	205 182
95	72.8 67.1	38.0 35.7	33.5 31.0	25.2 24.6	23.4 21.5	21.6 21.0	19.3 17.5	15.9 17.1	11.6 11.0	19.9 19.3	217 199

*H. W. Stoudt et al., *Weight, Height and Selected Body
Dimensions of Adults, United States 1960–1962*, Public
Health Service Publication no. 1000, ser. 11, no. 8.

Measuring and Weighing Test Subjects

In order to verify that one has representative test subjects (i.e., to represent established 5th and 95th percentiles), one should not depend entirely on the subjects' reports of their height or weight. In addition, in many instances the designer needs one or two other measurements that average subjects will not know about themselves. The rules for measuring are quite simple as long as a few rules are followed.

1. Stature

To measure stature, have the subject remove his or her shoes and stand with the head, shoulder blades, buttocks, and heels flush against a wall. Using a draftsman's triangle (see the accompanying illustration), measure the subject's height from the floor.

2. Arm Reach

Ask the subject, while he or she is still standing against the wall, to raise and point the arm straight forward with the palm of the hand facing inward and with the fingers extended. Measure the distance from the wall to the tip of the longest finger, usually the middle finger.

3. Leg Reach

Have the subject sit on a flat box with the head, shoulder blades, and buttocks against a wall. An adjustable seat and footstool should be used so that the subject's feet will be flat on the floor. Have the subject raise the right leg and lay it on the footstool. Older people may need help. Measure the distance from the wall to the subject's heel. A small triangle will help define this distance. Do not be disturbed by the fact that some people cannot fully straighten out their legs. However, have the subject straighten the leg as much as possible.

4. Weight

A standard health scale should be used. Subjects should be as lightly clothed as practical (see the accompanying illustration). This will require a nurse or another woman when the subject is a female.

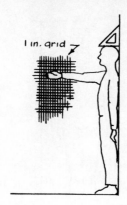

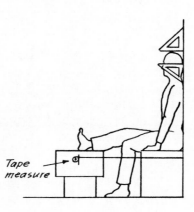

5. Strength

Although not absolutely valid, hand dynamometer strength is a relatively good indicator of an individual's overall strength capability. Hand dynamometers are not easy to find unless one is purchased for permanent retention. Such instruments often can be borrowed from a college or university that has an anthropology department. Usually one can also be found at a local hospital. Although instructions could be presented here, each dynamometer may be different, and thus one should ask for instructions from those from whom the particular instrument is borrowed.

The above measurements are usually sufficient for most design work where one is trying to make sure that test subjects are representative of a particular population's extremes. If, however, other measurements are required for some special design purpose, it is recommended that the designer seek help from a trained anthropologist.

STANDING HEIGHT (STATURE)*

Standing height is the primary indicator by which one selects general test subjects to evaluate various designs when there is no specific aspect of the design pertaining to an individual body component. The dimension is pertinent for adjusting head clearances.

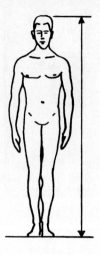

	Percentile		
	5th	50th	95th
Adults:			
Males	63.6 in	68.3 in	72.8 in
Females	59.0 in	62.9 in	67.1 in
Boys:			
Age 17	65.1 in	69.4 in	72.6 in
Age 14	58.0 in	63.2 in	69.1 in
Age 12	54.4 in	58.0 in	63.5 in
Age 6	41.8 in	45.1 in	49.1 in
Age 2	31.7 in	33.2 in	35.8 in
Girls:			
Age 17	60.0 in	64.1 in	67.6 in
Age 14	57.8 in	62.3 in	66.5 in
Age 12	54.1 in	58.8 in	63.0 in
Age 6	41.3 in	45.0 in	47.9 in
Age 2	30.4 in	33.0 in	35.4 in
Adults age 70 and over:			
Males	61.3 in	66.2 in	69.5 in
Females	55.3 in	61.8 in	64.9 in
Truck and bus drivers:			
Males	65.1 in	69.8 in	74.3 in
Females	58.9 in	63.5 in	68.0 in
Airline pilots (Male)	66.0 in	70.0 in	73.9 in
Flight attendants (Female)	62.5 in	65.4 in	68.8 in
Law enforcement officers:			
Males	66.6 in	70.0 in	74.0 in
Females	—	—	—

EYE HEIGHT (STANDING)

This dimension is pertinent to the location of visual displays and/or the sizing of visual obstructions, where a small person may have to see over the obstruction or over someone who is taller. As a general "rule of thumb," the eye height of males is about 5.2 in (13 cm) less than their standing height; for females it is about 4.8 in (12.2 cm) less. Similarly, when people stand normally (i.e., with some slump), their eye height lowers by about 1.6 in (4.1 cm) for males and by about 1.2 in (3.0 cm) for females.†

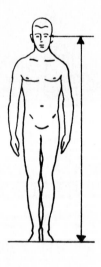

	Percentile		
	5th	50th	95th
Adults:			
Males	60.8 in	64.7 in	68.6 in
Females	57.3 in	60.3 in	65.3 in
Boys:			
Age 17	—	—	—
Age 14	—	—	—
Age 12	—	—	—
Age 6	—	—	—
Age 2	—	—	—
Girls:			
Age 17	—	—	—
Age 14	—	—	—
Age 12	—	—	—
Age 6	—	—	—
Age 2	—	—	—
Adults age 70 and over:			
Males	58.3 in	61.1 in	64.9 in
Females	50.8 in	55.8 in	60.8 in
Truck and bus drivers:			
Males	59.6 in	63.4 in	67.5 in
Females	—	—	—
Airline pilots (Male)	61.0 in	65.0 in	68.9 in
Flight attendants (Female)	58.0 in	60.9 in	64.3 in
Law enforcement officers:			
Males	61.6 in	65.0 in	69.0 in
Females	—	—	—

*The data in this table and the following tables have been compiled from the several references cited at the end of this section.

†Rules of thumb apply only to adults.

OVERHEAD REACH (STANDING)

This dimension is pertinent to locating controls that are overhead. It should be used in conjunction with stature because, although a short person must be able to reach a control, it should not be so low that it becomes an obstruction for the taller person. Use this dimension to select subjects for evaluating the accessibility of objects on high shelves.

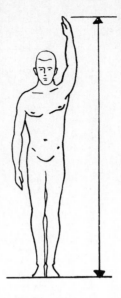

	Percentile		
	5th	50th	95th
Adults:			
Males*	82.0 in	88.0 in	94.0 in
Females*	73.0 in	79.0 in	86.0 in
Boys:			
Age 17	77.7 in	84.1 in	88.6 in
Age 14	70.5 in	77.3 in	83.1 in
Age 12	64.7 in	70.7 in	78.9 in
Age 6	50.3 in	53.3 in	58.2 in
Age 2	38.2 in	41.7 in	48.1 in
Girls:			
Age 17	72.6 in	77.7 in	81.9 in
Age 14	66.1 in	75.2 in	83.1 in
Age 12	65.0 in	72.1 in	77.4 in
Age 6	48.1 in	53.0 in	56.8 in
Age 2	38.7 in	41.6 in	45.6 in
Adults age 70 and over:			
Males	—	—	—
Females	69.5 in	75.5 in	82.5 in
Truck and bus drivers:			
Males*	81.0 in	88.1 in	93.7 in
Females	—	—	—
Airline pilots* (Male)	85.0 in	91.5 in	95.1 in
Flight attendants* (Female)	65.8 in	68.9 in	72.3 in
Law enforcement officers:			
Males*	85.0 in	89.7 in	95.2 in
Females	—	—	—

*Estimated.

FORWARD REACH (STANDING)

This dimension is pertinent to the selection of test subjects who will be evaluating reach conditions within the design situation. This dimension should be used in conjunction with leg-reach dimensions when the design situation calls for a seated operator to operate both hand and foot controls.

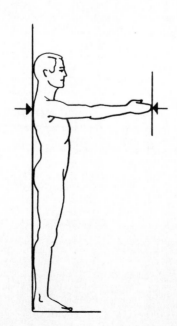

	Percentile		
	5th	50th	95th
Adults:			
Males	31.9 in	34.6 in	37.3 in
Females	29.7 in	31.8 in	34.1 in
Boys:			
Age 17	25.7 in	28.7 in	32.0 in
Age 14	23.8 in	26.9 in	29.3 in
Age 12	22.2 in	24.2 in	26.8 in
Age 6	17.6 in	19.1 in	20.9 in
Age 2	14.0 in	15.8 in	18.1 in
Girls:			
Age 17	23.9 in	26.7 in	28.8 in
Age 14	22.8 in	25.5 in	28.4 in
Age 12	21.8 in	24.3 in	26.4 in
Age 6	16.6 in	18.7 in	20.6 in
Age 2	13.7 in	15.1 in	16.8 in
Adults age 70 and over:			
Males	—	33.5 in	—
Females	—	—	—
Truck and bus drivers:			
Males	33.0 in	35.8 in	38.4 in
Females	—	—	—
Airline pilots (Male)	32.9 in	34.3 in	37.0 in
Flight attendants (Female)	29.0 in	31.0 in	33.3 in
Law enforcement officers:			
Males	32.3 in	34.8 in	37.7 in
Females	—	—	—

BUTTOCK-TO-POPLITEAL LENGTH

This dimension is pertinent to seat length. Although it is desirable to provide adequate support for the larger person, it is the shorter person who will have the most problems if this dimension is ignored in the test subject sample.

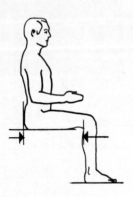

	Percentile		
	5th	50th	95th
Adults:			
Males	17.3 in	19.5 in	21.6 in
Females	17.0 in	18.9 in	21.0 in
Boys:			
Age 17			
Age 14			
Age 12			
Age 6			
Age 2			
Girls:			
Age 17			
Age 14			
Age 12			
Age 6			
Age 2			
Adults age 70 and over:			
Males	17.0 in	18.9 in	21.2 in
Females	17.0 in	18.7 in	20.0 in
Truck and bus drivers:			
Males	20.2 in	22.2 in	24.0 in
Females	18.4 in	20.5 in	22.3 in
Airline pilots			
Flight attendants (Female)	17.4 in	19.0 in	20.6 in
Law enforcement officers:			
Males	—	—	—
Females	—	—	—

MIDSHOULDER HEIGHT (SITTING)

This dimension is pertinent to the location of controls (i.e., most controls should not be located above the shoulder) and to the location of the anchor point for seat belts (i.e., the belt should not depart aft of the shoulder horizontally or at a negative angle; otherwise, it will create discomfort and tend to push the occupant downward during a frontal-impact crash).

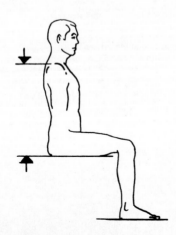

	Percentile		
	5th	50th	95th
Adults:			
Males	21.0 in	—	25.0 in
Females	18.0 in	—	25.0 in
Boys:			
Age 17	—	—	—
Age 14	18.1 in	20.1 in	21.2 in
Age 12	17.8 in	19.4 in	21.5 in
Age 6	13.7 in	15.4 in	17.0 in
Age 2	11.1 in	12.3 in	13.3 in
Girls:			
Age 17	—	—	—
Age 14	19.1 in	21.2 in	22.6 in
Age 12	17.6 in	19.5 in	21.0 in
Age 6	13.6 in	15.1 in	16.5 in
Age 2	11.1 in	12.3 in	14.0 in
Adults age 70 and over:			
Males	20.9 in	24.0 in	27.2 in
Females	—	—	—
Truck and bus drivers:			
Males	21.2 in	24.3 in	27.5 in
Females	—	—	—
Airline pilots	—	—	—
Flight attendants (Female)	20.1 in	22.9 in	25.7 in
Law enforcement officers:			
Males	—	—	—
Females	—	—	—

SITTING HEIGHT

This dimension is pertinent to the establishment of proper overhead clearances for seated persons. It is particularly important in the design of driver work stations.

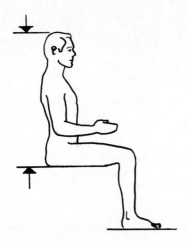

	Percentile		
	5th	50th	95th
Adults:			
Males	33.2 in	35.7 in	38.0 in
Females	30.9 in	33.4 in	35.7 in
Boys:			
Age 17	33.7 in	35.9 in	37.8 in
Age 14	29.4 in	32.3 in	35.6 in
Age 12	28.2 in	29.9 in	32.5 in
Age 6	23.0 in	25.0 in	26.8 in
Age 2	19.0 in	20.6 in	21.9 in
Girls:			
Age 17	31.8 in	33.7 in	35.9 in
Age 14	29.9 in	32.4 in	34.8 in
Age 12	28.0 in	30.4 in	32.9 in
Age 6	22.8 in	24.9 in	26.5 in
Age 2	18.9 in	19.8 in	21.5 in
Adults age 70 and over:			
Males	31.8 in	34.3 in	36.7 in
Females	28.1 in	32.1 in	34.8 in
Truck and bus drivers:			
Males	34.3 in	36.3 in	38.2 in
Females	—	—	—
Airline pilots (Male)	—	—	—
Flight attendants (Female)	32.4 in	34.3 in	36.1 in
Law enforcement officers:			
Males	34.1 in	36.3 in	38.5 in
Females	—	—	—

EYE HEIGHT (SITTING)

This dimension is pertinent to the design of work stations in which visual displays and/or outside viewing requires accommodation of a range of operator sizes. As a rule of thumb the eye height of males is about 5.2 inch (13 cm) less than their sitting height; that of females is about 4.8 in (12.2 cm) less. Similarly, when people sit normally (with some slump), their eye height lowers by about 1.2 in (3.0 cm) for males and by about 1.2 in (3.0 cm) for females.

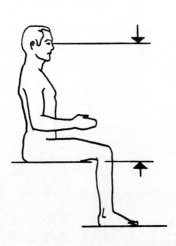

	Percentile		
	5th	50th	95th
Adults:			
Males	28.7 in	31.3 in	33.5 in
Females	27.4 in	29.3 in	31.0 in
Boys:			
Age 17	29.7 in	31.6 in	33.2 in
Age 14	25.2 in	28.2 in	30.7 in
Age 12	23.6 in	25.4 in	27.0 in
Age 6	18.3 in	20.5 in	22.2 in
Age 2	15.9 in	17.6 in	18.8 in
Girls:			
Age 17	26.9 in	29.0 in	31.1 in
Age 14	25.1 in	28.0 in	30.4 in
Age 12	23.6 in	25.6 in	27.6 in
Age 6	17.8 in	20.6 in	22.4 in
Age 2	15.2 in	16.7 in	18.1 in
Adults age 70 and over:			
Males	—	26.8 in	—
Females	—	—	—
Truck and bus drivers:			
Males	27.7 in	29.6 in	31.6 in
Females	—	—	—
Airline pilots (Male)	29.4 in	31.5 in	33.5 in
Flight attendants (Female)	28.1 in	29.9 in	31.7 in
Law enforcement officers:			
Males	29.1 in	31.3 in	33.5 in
Females	—	—	—

BUTTOCK-TO-KNEE LENGTH

This dimension is pertinent to establishing knee clearance for the seated operator. It should be used in conjunction with knee-height and thigh-clearance dimensions.

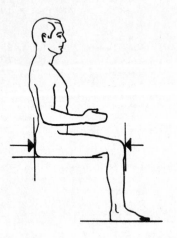

	Percentile		
	5th	50th	95th
Adults:			
Males	21.3 in	23.3 in	25.2 in
Females	20.4 in	22.4 in	24.6 in
Boys:			
Age 17	21.2 in	23.1 in	25.0 in
Age 14	19.2 in	21.4 in	23.5 in
Age 12	17.7 in	19.5 in	21.5 in
Age 6	12.9 in	14.1 in	15.6 in
Age 2	8.5 in	9.3 in	10.5 in
Girls:			
Age 17	20.0 in	21.7 in	23.5 in
Age 14	19.4 in	21.2 in	23.2 in
Age 12	17.9 in	20.1 in	22.0 in
Age 6	12.5 in	14.2 in	15.4 in
Age 2	7.4 in	9.4 in	10.6 in
Adults age 70 and over:			
Males	21.0 in	22.6 in	24.4 in
Females	19.9 in	22.2 in	23.9 in
Truck and bus drivers:			
Males	22.7 in	24.6 in	26.8 in
Females	20.6 in	22.9 in	24.9 in
Airline pilots (Male)	22.0 in	23.6 in	25.6 in
Flight attendants (Female)	21.2 in	22.6 in	24.2 in
Law enforcement officers:			
Males	22.6 in	24.2 in	26.1 in
Females	—	—	—

POPLITEAL HEIGHT (SITTING)

This dimension is pertinent to the establishment of appropriate seat heights. It is also pertinent to the selection of test subjects who will be used to evaluate the relationships between a vehicle seat and foot controls.

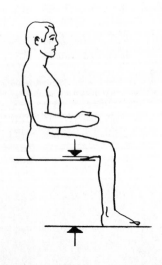

	Percentile		
	5th	50th	95th
Adults:			
Males	15.5 in	17.3 in	19.3 in
Females	14.0 in	15.7 in	17.5 in
Boys:			
Age 17	—	—	—
Age 14	—	—	—
Age 12	13.2 in	14.6 in	16.1 in
Age 6	10.4 in	11.5 in	12.6 in
Age 2	—	—	—
Girls:			
Age 17	—	—	—
Age 14	—	—	—
Age 12	13.0 in	14.7 in	16.3 in
Age 6	10.2 in	11.3 in	12.5 in
Age 2	—	—	—
Adults age 70 and over:			
Males	15.2 in	16.6 in	17.9 in
Females	13.5 in	15.6 in	17.2 in
Truck and bus drivers:			
Males	15.7 in	17.5 in	19.7 in
Females	—	—	—
Airline pilots (Male)	15.7 in	17.0 in	18.2 in
Flight attendants (Female)	15.9 in	17.1 in	18.5 in
Law enforcement officers:			
Males	—	—	—
Females	—	—	—

KNEE HEIGHT (SITTING)

This dimension is pertinent to the establishment of knee clearance. It should be used in conjunction with buttock-to-knee length and thigh-clearance dimensions.

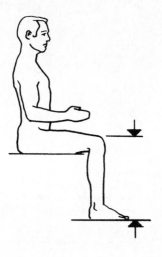

	Percentile		
	5th	50th	95th
Adults:			
Males	19.3 in	21.4 in	23.4 in
Females	17.9 in	19.6 in	21.5 in
Boys:			
Age 17	19.4 in	21.8 in	23.4 in
Age 14	18.5 in	20.3 in	22.6 in
Age 12	17.2 in	18.6 in	20.7 in
Age 6	12.4 in	13.9 in	15.3 in
Age 2	8.1 in	9.4 in	10.2 in
Girls:			
Age 17	17.8 in	19.5 in	21.2 in
Age 14	17.7 in	19.5 in	21.1 in
Age 12	17.0 in	18.6 in	20.2 in
Age 6	12.4 in	13.5 in	14.8 in
Age 2	8.0 in	9.4 in	10.0 in
Adults age 70 and over:			
Males	19.0 in	20.7 in	22.7 in
Females	17.3 in	19.4 in	20.9 in
Truck and bus drivers:			
Males	20.1 in	21.7 in	23.5 in
Females	—	—	—
Airline pilots (Male)	20.1 in	21.7 in	23.3 in
Flight attendants (Female)	19.1 in	20.4 in	21.9 in
Law enforcement officers:			
Males	20.4 in	22.0 in	23.7 in
Females	—	—	—

THIGH CLEARANCE (SITTING)

This dimension is pertinent to the establishment of the proper clearance between a seat and the lower edge of a desk and to the establishment of the space between the driver's seat and a vehicle steering wheel.

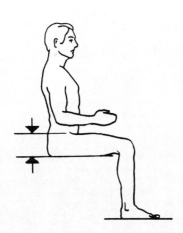

	Percentile		
	5th	50th	95th
Adults:			
Males	4.3 in	5.7 in	6.9 in
Females	4.1 in	5.4 in	6.9 in
Boys:			
Age 17	4.8 in	6.1 in	6.9 in
Age 14	4.3 in	5.5 in	6.4 in
Age 12	4.2 in	5.0 in	5.8 in
Age 6	3.1 in	3.6 in	4.3 in
Age 2	2.5 in	3.3 in	3.9 in
Girls:			
Age 17	4.6 in	5.4 in	6.8 in
Age 14	4.6 in	5.4 in	6.1 in
Age 12	4.2 in	4.9 in	6.1 in
Age 6	3.0 in	3.7 in	4.5 in
Age 2	2.1 in	3.1 in	3.7 in
Adults age 70 and over:			
Males	4.1 in	5.2 in	6.6 in
Females	4.0 in	5.2 in	6.5 in
Truck and bus drivers:			
Males	5.0 in	5.9 in	6.9 in
Females	—	—	—
Airline pilots (Male)	4.8 in	5.6 in	6.5 in
Flight attendants (Female)	—	—	—
Law enforcement officers:			
Males	—	—	—
Females	—	—	—

ELBOW-TO-FINGERTIP LENGTH

This dimension is pertinent to special problems such as evaluating whether a control may have been placed too close to the operator, with the result that (because of the seat) the operator cannot get his or her arm back far enough to use the control properly.

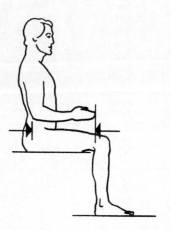

	Percentile		
	5th	50th	95th
Adults:			
Males	—	—	—
Females	—	—	—
Boys:			
Age 17	17.5 in	18.7 in	20.0 in
Age 14	15.6 in	17.3 in	19.1 in
Age 12	14.4 in	15.6 in	17.3 in
Age 6	10.9 in	12.0 in	13.2 in
Age 2	7.6 in	8.9 in	9.9 in
Girls:			
Age 17	15.5 in	16.8 in	18.0 in
Age 14	15.2 in	16.5 in	17.9 in
Age 12	14.0 in	15.7 in	17.3 in
Age 6	10.7 in	11.7 in	12.8 in
Age 2	7.7 in	8.8 in	9.4 in
Adults age 70 and over:			
Males	17.2 in	18.3 in	19.5 in
Females	—	—	—
Truck and bus drivers:			
Males	17.4 in	18.8 in	20.4 in
Females	15.1 in	16.7 in	18.2 in
Airline pilots (Male)	17.6 in	18.9 in	20.2 in
Flight attendants (Female)	16.0 in	17.2 in	18.3 in
Law enforcement officers:			
Males	—	—	—
Females	—	—	—

WAIST DEPTH

This dimension is pertinent to the establishment of clearance requirements between the backrest of a chair or seat and the leading edge of a worktable, desk, or steering wheel.

	Percentile		
	5th	50th	95th
Adults:			
Males	7.1 in	9.7 in	12.3 in
Females	5.8 in	6.6 in	7.9 in*
Boys:			
Age 17			
Age 14			
Age 12			
Age 6			
Age 2			
Girls:			
Age 17			
Age 14			
Age 12			
Age 6			
Age 2			
Adults age 70 and over:			
Males			
Females			
Truck and bus drivers:			
Males			
Females			
Airline pilots (Male)	7.6 in	9.9 in	10.4 in
Flight attendants (Female)	5.2 in	5.8 in	6.5 in
Law enforcement officers:			
Males			
Females			

*In certain working situations, one should consider the additional requirement for clearance when the pregnant woman must be accommodated. Unfortunately, to our knowledge, there are no data on this additional waist-clearance factor.

ELBOW REST HEIGHT

Not only is this dimension pertinent to the establishment of armrest heights, but it also provides a basis for establishing the level of a writing surface and/or the approximate position of the middle row of a keyboard, the location of a joy-stick handle or control wheel, etc.

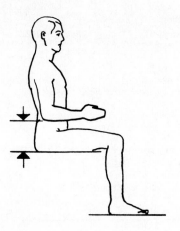

	Percentile		
	5th	50th	95th
Adults:			
Males	7.4 in	9.5 in	11.6 in
Females	7.1 in	9.2 in	11.0 in
Boys:			
Age 17			
Age 14			
Age 12			
Age 6			
Age 2			
Girls:			
Age 17			
Age 14			
Age 12			
Age 6			
Age 2			
Adults age 70 and over:			
Males	6.5 in	8.6 in	10.6 in
Females	6.4 in	8.4 in	10.0 in
Truck and bus drivers			
Males	—	—	—
Females	—	—	—
Airline pilots (Male)	7.4 in	9.1 in	10.8 in
Flight attendants (Female)	7.7 in	9.6 in	11.0 in
Law enforcement officers:			
Males	—	—	—
Females	—	—	—

SHOULDER BREADTH

This dimension is pertinent to the establishment of lateral clearance between persons who may be required to sit side by side and to the establishment of the lateral clearance requirement for a worker who may have to squeeze into a tight space to work on an item of equipment.

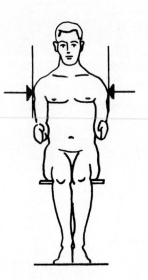

	Percentile		
	5th	50th	95th
Adults:			
Males	16.4 in	17.9 in	19.6 in
Females	14.4 in	15.7 in	17.6 in
Boys:			
Age 17	15.5 in	17.4 in	19.2 in
Age 14	13.7 in	15.3 in	17.4 in
Age 12	12.3 in	13.8 in	15.6 in
Age 6	9.3 in	11.0 in	12.5 in
Age 2	8.0 in	8.9 in	9.8 in
Girls:			
Age 17	14.2 in	15.6 in	17.3 in
Age 14	13.4 in	14.8 in	16.7 in
Age 12	12.2 in	13.8 in	15.8 in
Age 6	10.0 in	10.9 in	11.9 in
Age 2	7.6 in	9.0 in	9.7 in
Adults Age 70 and over:			
Males	15.6 in	17.0 in	18.5 in
Females	—	—	—
Truck and bus drivers:			
Males	16.9 in	18.3 in	19.9 in
Females	—	—	—
Airline pilots (Male)	16.5 in	17.9 in	19.4 in
Flight attendants (Female)	14.9 in	16.0 in	17.0 in
Law enforcement officers:			
Males	17.6 in	19.4 in	21.4 in
Females	—	—	—

HIP BREADTH (SITTING)

This dimension is pertinent to the establishment of seat widths, particularly the lateral clearance between armrests.

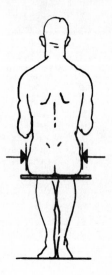

	Percentile		
	5th	50th	95th
Adults:			
Males	12.2 in	14.0 in	15.9 in
Females	12.3 in	14.3 in	17.1 in
Boys:			
Age 17	11.9 in	13.1 in	14.8 in
Age 14	10.4 in	11.8 in	13.6 in
Age 12	9.2 in	10.7 in	12.8 in
Age 6	7.5 in	8.3 in	9.4 in
Age 2	6.9 in	7.5 in	8.4 in
Girls:			
Age 17	12.0 in	13.5 in	15.7 in
Age 14	11.2 in	12.7 in	14.4 in
Age 12	9.6 in	11.3 in	13.5 in
Age 6	7.4 in	8.4 in	9.3 in
Age 2	6.6 in	7.4 in	8.4 in
Adults age 70 and over:			
Males	12.1 in	13.6 in	15.5 in
Females	11.7 in	14.0 in	16.8 in
Truck and bus drivers:			
Males	13.2 in	14.5 in	16.3 in
Females	—	—	—
Airline pilots (Male)	12.1 in	13.2 in	14.4 in
Flight attendants (Female)	13.3 in	14.5 in	15.6 in
Law enforcement officers:			
Males	—	—	—
Females	—	—	—

FOREARM-TO-FOREARM BREADTH

This dimension is pertinent to the establishment of seat and armrest separation. It also should be considered along with shoulder breadth to establish lateral body clearance for side-by-side seating.

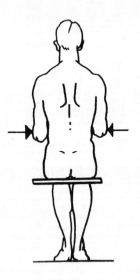

	Percentile		
	5th	50th	95th
Adults:			
Males	13.7 in	16.5 in	19.9 in
Females	12.3 in	15.1 in	19.3 in
Boys:			
Age 17			
Age 14			
Age 12			
Age 6			
Age 2			
Girls:			
Age 17			
Age 14			
Age 12			
Age 6			
Age 2			
Adults age 70 and over:			
Males	14.0 in	16.4 in	18.7 in
Females	13.1 in	15.7 in	19.1 in
Truck and bus drivers:			
Males	16.6 in	19.4 in	23.1 in
Females	14.1 in	16.5 in	20.2 in
Airline pilots (Male)	15.2 in	17.2 in	19.8 in
Flight attendants (Female)	11.6 in	13.0 in	14.6 in
Law enforcement officers:			
Males	—	—	—
Females	—	—	—

CORRELATION AMONG BODY DIMENSIONS*

Some body dimensions correlate rather well with others, making it possible to select test subjects for special design evaluation activities.

However, although ordinarily one would expect tall persons to have long legs, arms, and torsos, and short persons to have short legs, arms, and torsos, to ensure that a particular design does not cause problems for special members of the expected user population, it is important not to make too many assumptions about dimensional correlations. The accompanying table presents those correlations that indicate safe assumptions and those that do not.

In addition, consider the following suggestions:

For reach problems select a variety of subjects who variously have short stature, short legs, and short arms.

For clearance problems select a variety of subjects who variously are tall, have long legs and/or sitting heights, and have wide shoulders and hip breadths.

If the weight of a subject is pertinent (in addition to the other dimensional characteristics noted above), consider, for example, one or two heavy but small subjects, for they may sink into a seat and spoil your plans for stabilizing a range of eye references or head clearances.

Dimension	Stature	Chest Circumference
Weight		0.820
Torso		
Sitting height	0.722	
Trunk height	0.587	
Cervical height	0.948	
Waist circumference		0.745
Hip circumference		0.744
Bideltoid		0.706
Cross-back width		0.486
Hip breadth		0.649
Shoulder circumference		0.806
Neck circumference		0.627
Arm		
Inside arm length	0.718	
Sleeve length	0.624	
Inseam	0.819	
Forearm-hand length	0.640	
Shoulder-elbow length	0.660	
Elbow breadth		0.677
Wrist circumference		0.556
Leg		
Lower leg	0.688	
Patella height	0.795	
Buttock-knee height	0.751	
Outseam	0.886	
Total crotch length		0.467
Crotch-thigh circumference		0.731

EFFECTS OF CLOTHING ON HUMAN BODY DIMENSIONS†

As noted previously, most human body dimensions were obtained using nude subjects. For practical applications, therefore, one must consider the effect that clothing, shoes, gloves, hats, or other items of apparel may have on specific base-line dimensions. The accompanying table provides rough approximations of how much clothing adds to, or subtracts from, given body dimensions. However, when a particular design problem indicates the possibility that some particular clothing item may influence clearance, reach, etc., one should select an appropriate subject sample and have subjects put on the specific clothing items pertinent to the expected user environment. Remember that clothing not only increases clearance dimensions but also often creates considerable restrictions to mobility and reach capability.

Key questions with regard to what type of clothing should be considered include whether an operation will be performed in shirt sleeves, at low temperatures, under reduced pressure, or in other special environments.

Effect of Various Types of Clothing on Human Body Dimensions

Dimension	Street Clothes Men	Street Clothes Women	Winter Clothes Men	Winter Clothes Women	Heavy Flight Clothing	Pressure Suits* Unpressurized	Pressure Suits* Pressurized
Weight	5 lb	3½ lb	10 lb	7 lb	12–15 lb	21 lb	21 lb
Stature	1 in	½–3¾ in	1 in	½–3¾ in	3 in	3½ in	2½ in
Vertical reach	1	½–3¾	1	½–3¾	1	(−2½)	(−16½)
Eye height, standing	1	½–3¾	1	½–3¾	1	(−3½)	2½
Crotch to floor	1	½–3¾	1	½–3¾	1	(−1)	(−1)
Foot length	1¼	½	1½	½–¾	1	1	1
Foot width	½–¾	¼–(−½)†	½–1	¼–½	¾	¾	¾
Head length	—	—	—‡		4½	4½	4½
Head width	—	—	—‡		4½	4½	4½
Hand length	—	—	¾	½	½	½	¼
Hand width	—	—	½	¼	½	½	1
Hand thickness	—	—	½	¼	½	¾	1¼
Fist circumference	—	—	1	¾	1	1¼	3
Shoulder width	½	¼	2–3	1	1½	1	½
Hip width	½	¼	2–3	1	1½	1	2¾
Elbow-to-elbow width	¾	¼	2–3½	1–1½	1	6	9
Thigh clearance	½	¼	1	¾	2	1¾	2
Forearm-to-fist length	½	¼	¾	½	1	1½	5½

*Certain pressurized garments are designed for seated position and therefore shorten certain dimensions. The helmet, however, tends to rise under pressure.

†Women's dress shoes confine and shrink foot width.

‡An Army steel helmet is approximately 12 by 10¼ in.

*W. E. Woodson and D. W. Conover, *Human Engineering Guide for Equipment Designers,* University of California Press, Berkeley, 1964.

†W. E. Woodson and D. W. Conover, *Human Engineering Guide for Equipment Designers,* University of California Press, Berkeley, 1964.

SPECIAL TOOLS

Two-Dimensional Drafting Templates

Two-dimensional templates representing the physical dimensions of the human body are useful for initial estimates of operator-equipment interface "fit." Although some templates are available (see the accompanying illustration), one generally has to fabricate his or her own set of manikins to whatever scale is desired. To accomplish the typical range of drawing applications satisfactorily, one should develop and use both male and female templates in both the 95th- and the 5th-percentile ranges. The 5th percentile is needed to evaluate reach problems, and the 95th to evaluate clearance problems. Both will be required to evaluate eye-level reference requirements, since both tall and short people must be accommodated.

Two-dimensional drafting templates designed for automotive use can be ordered from the Society of Automotive Engineers, 400 Commonwealth Drive, Warrendale, PA 15096.

Two-dimensional templates representing U.S. Air Force personnel can be ordered from the 6570th Aerospace Medical Research Laboratory, Wright-Patterson Air Force Base, Ohio 45433, Attention: Mr. Kenneth W. Kennedy.

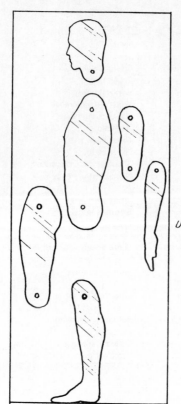

Unassembled

The accompanying illustrations show how an articulated manikin is cut out and joined together to provide an articulating template for drawing around. It is suggested that manikin parts be made of (at least) 1/16-in (0.16-cm) transparent plastic (preferably fluorescent orange) so that one can see through the template while drawing.

Following are specifications for three of the most often-used templates, i.e., a 5th-percentile male, 95th-percentile adult male, and a 5th-percentile adult female. Note that these templates are representative of nude dimensions, and thus clothing allowances should be made during design.

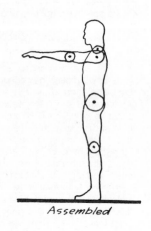

Assembled

5th-Percentile Civilian Male Manikin

Manikins can be made out of transparent plastic. Parts should be riveted so that they articulate freely. The thickness of the plastic should be sufficient to provide adequate rigidity for ease of tracing around the manikin (the optimum thickness will be a function of the selected scale and overall size of the finished manikin).

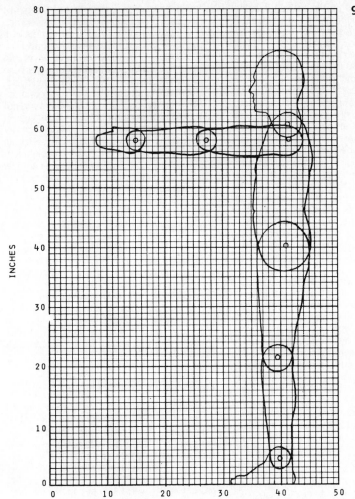

95th-Percentile Civilian Male Manikin

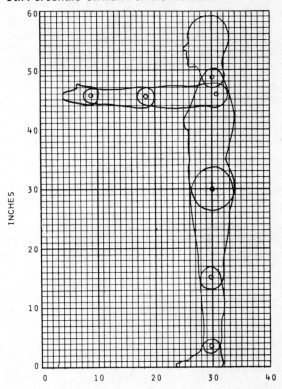

5th-Percentile Civilian Female Manikin

TESTING

General Considerations for Conducting Human Engineering Tests

First, it is highly recommended that an experienced human factors specialist be employed to assist in the design and conduct of human engineering tests whenever this is practicable. It is particularly important if the proposed test deals with several highly interactive variables that require complex experimental designs in order to isolate and establish the significance of certain variables relative to human performance. It is equally important to acquire help when a particular test may involve possible hazards to the test subjects.

However, designers can conduct certain types of tests on their own as long as they keep in mind certain problems associated with testing that involves human subjects. The following should be considered:

1. A sufficient number of test subjects and trials should be used to ensure that the test results do not represent some sporadic incident not representative of the true performance of the subjects.
2. Sufficient procedural control should be provided so that one can safely infer that, "when this adjustment or input occurs, with these conditions, this is the typical response or human output." Generally one must keep certain features constant while a single variable is adjusted so that, in the end, it is possible to correlate specific conditions with specific human responses.
3. In most cases, test subjects should not be privy to the real purpose or objectives of the test; they should be informed only of the procedures they are to follow. This helps avoid "expectancy biases," wherein subjects may attempt to give the experimenter the response they think the experimenter wants, rather than performing according to their individual or unique response characteristics.
4. The test procedure should cause entry and sequential aspects of the test to be randomly presented; i.e., it should not be possible for the subject to "learn" a sequence or anticipate a step because of start cues. Attempt to remove any possibility that "order effects" show up in the final subject response.
5. When a particular hardware configuration is being evaluated in terms of whether subjects perform better on one version than another, make sure that alternative configurations are equally well represented (in terms of quality of realism, workmanship, and to visual attractiveness).
6. Experimenters should remain as nonconspicuous as possible during evaluations in order to minimize inadvertent cues they might provide the test subjects during a test sequence.

7. Attempt to set up the test so that the trials are long enough to identify the difference between "learning" and informed response. However, trials should not be so long that fatigue produces artifacts in the results that have little to do with the basic evaluation objectives. Note, however, that fatigue effects may be a significant test objective, in which case one must design the test so as to differentiate between design-induced fatigue and fatigue that results merely from the fact that the test subjects are generally tired or bored with the activity.
8. Whenever possible, design all tests in such a way that quantitative (as well as qualitative) results can be documented. Avoid general opinion-oriented evaluations unless these are quantified in terms of forced choice ratings or rankings by test subjects. Incidentally, this is desirable even for general VIP evaluations of a proposed design. Although some visitors may object to being used as test subjects, it is possible to formalize their review in a manner which is not objectionable to them but which is organized sufficiently to allow one to tabulate and compare opinions.
9. Select the right subjects; e.g., avoid using engineers to represent a typical nonengineering user population.
10. Obtain important demographic information about each subject so that, in the event it becomes necessary to evaluate test results in terms of subject differences, you do not have to try to find a particular subject again. In fact, attempt to screen out subjects who for some reason may not be able to complete a test series or who may not be available for a second test.
11. When a test involves visual acuity (visual display, illumination level, etc.), either select subjects on the basis of their having normal acuity (by giving the subjects an eye test), or specifically seek out and include subjects who have the typical visual anomalies you expect among the eventual user population. For example, if the user population will include older people, who normally require higher illumination levels and larger visual detail to see properly, be sure to include subjects who are older and may have to wear corrective lenses.
12. When the design to be tested may be used by persons with various handicaps, select a broad range of handicapped subjects because, although a given subject may have a particular type of handicap, each such individual tends to vary in terms of how he or she copes with various mobility, manipulative, visual, or auditory tasks.
13. Carefully plan any human engineering task in terms of the procedures to be used and practice conducting the test prior to conducting the main test. Re-

member that, once a test has begun, it should be continued to completion; i.e., avoid having to stop in the middle of a test sequence because you have forgotten something or have lost your place in the test sequence or because the data-recording system has failed. Nothing aggravates test subjects more than feeling that the experimenter is not sure of what he or she is doing.
14. If a particular design evaluation involves special considerations, such as time of day (i.e., both day and night ambient illumination conditions) or special clothing (i.e., light summer clothing and/or heavy winter clothing), be sure to include these conditions in the test.
15. Make sure that the mockup and/or test apparatus is fully checked out and will not fail during the test. Be especially careful to see that there are no unforeseen hazards either for the subjects or for the experimenters. If there appear to be certain stress conditions associated with the test, make sure that none of the subjects who are selected have tendencies or conditions that could be compromised during the stressing phases of the test (a potential for epileptic seizure, heart failure, etc.).

Field Testing and Evaluation

Nothing provides a more conclusive demonstration of adequate design than the actual field test of a prototype model. Typically this is done in the case of airplanes, highway vehicles, seacraft, agricultural machines, etc. Too often, however, such tests rely on the opinions of so-called expert test personnel, e.g., test pilots or test drivers. Although there is often a good reason to utilize these experts (for reasons of safety and because they are perhaps more skilled and perceptive in their analysis of features that may not be quite right), the real proof of design acceptability comes when the intended user (who lacks professional skill) can also operate the system.

Although the main objective of most field tests is to prove that the system will do what it is supposed to do, too often one may not obtain the type of information from such tests to suggest modifications that will make the system easier to operate by the average user. In addition, the field test should provide other information that is important to the eventual user; e.g., previously described tasks can be verified, training objectives can be confirmed, and training aids can be evaluated.

A field test should be designed with the following objectives in mind:

It will demonstrate the reliability of the hardware under all mission and environmental conditions.

It will demonstrate that the training program provided all the necessary skills for the operator or maintainer to cope with the operational conditions under which the system is to be used.

It will pinpoint any hardware, software, operator interface, or procedural discrepancies in such a way that one knows what to do to correct deficiencies in any of these areas (e.g., hardware, software, human interface, procedures, and training).

It will allow one to verify whether estimates of manning level are correct.

It will allow one to evaluate whether estimates of time factors were correct (service turn-around, loading, boarding, emergency escape, etc.).

It will help determine any safety hazards that may not have been anticipated.

It will provide an initial impression of consumer acceptability of the product.

Methods and Techniques for Human Engineering Field Testing

OPERATOR PERFORMANCE MEASUREMENT A quantitative measure of operator and maintainer performance in terms of time and error should be provided; i.e., in addition to the probable effort to elicit personal evaluations by the test subjects and/or expert test operators, actual measurement of the performance of these participants should always be generated wherever practicable.

COMPLETE TEST OPERATIONAL SCENARIO The field test should be designed and implemented so that one systematically exercises the human-machine system through all phases of the operating scenario. This should include both normal and emergency conditions. The scenario should also be complete in terms of all the human-system interactions: visual, auditory, mobility, dexterity, communications, decision making, and control.

REAL-WORLD ENVIRONMENT The test action should occur on site, i.e., on real roadways and streets, in rough water, during the day and at night, and under different environmental conditions (rain, snow, heat, cold, high wind, fog, etc.).

PERFORMANCE RECORDING In addition to the actual instrumentation of operator interface responses noted above, consider the use of other covert techniques such as voice recorders, motion picture cameras, and human physiological sensors (pulse, heart rate, galvanic skin resistance, etc.).

OPERATOR OPINION Provide a means for documenting the opinions of test pilots, drivers, and/or typical test subjects, not only in terms of posttest debriefing, but also in terms of on-line evaluation during the testing (e.g., a voice recorder). Debriefing questionnaires should be designed to elicit specific, design-related comments, as opposed to general verbal descriptions of what the subject felt about the equipment.

TEST OBSERVERS Where appropriate, consider the use of trained independent observers to take notes while the test is under way. The independent observer may be important because the actual test subject is sometimes too busy to observe critical events as they are happening.

Trained observers are also effective in performing general population activities. For example, strategically located observers can observe and document how many drivers are wearing seat belts, how pedestrians behave with respect to crossing streets or using paths on a campus, or how individuals utilize a particular piece of equipment, such as a ticket vendor, a coin-operated laundry machine, an elevator, or an escalator.

Equip observers with tabulating devices where appropriate (stopwatch, hand counter, sketch pad, questionnaire and documentation record pad, etc.).

TEST SUBJECT INDOCTRINATION All field test subjects should be properly indoctrinated before they participate in any field testing. Depending on the complexity of the test, try to develop a brief statement of the purpose and objectives of the test and a general overview of what will be expected of the subject prior to, during, and following the test. This should be written down so that all subjects receive the same instructions. In some cases, it may also be desirable to provide written instructions during the test; i.e., so that each subject receives exactly the same instructions, each step is prepared and presented on a slide or by means of a recording or is read to the subject by the experimenter.

TEST CONDUCT Although it is generally best to leave the subjects alone as much as possible in order to reduce the possible influence of an experimenter's method of instruction, voice inflection, or body language, it is sometimes desirable to have the subjects verbalize about what they are doing while the test is in progress. This should be part of the indoctrination, with possible reminders by the experimenter if the subjects seem to forget.

When several different experimenters are to be used, allow them to practice until they demonstrate that they perform consistently and in a similar fashion.

In certain kinds of field testing, it will be necessary to assign the responsibility for safety to a given individual; i.e., although each experimenter is responsible for the safety of his or her particular subject during the test, general overall safety monitoring should be provided. This is especially important when several equipment or vehicle components are engaged in an interactive operation.

Safety equipment should be provided to deal with emergencies in certain types of field tests, such as fire trucks, rescue equipment and personnel, and medical personnel.

When extended-duration field tests are being conducted, it may be necessary to provide housing, food, or other support for both the test personnel and the test subjects.

IN-THE-FIELD TEST DATA ANALYSIS Depending on the nature, location, and duration of a particular field test, it may be important to provide facilities and equipment for analyzing data on the spot. Such data may be important to the test conductor in terms of deciding whether to make modifications to the test schedule or procedures. Alternatively, it may be possible to transfer data via various telecommunication methods to a base site or to a laboratory where large computer facilities are available. Such remote analysis is often required when the basic system is operating in the air or in space.

HAZARDOUS-ENVIRONMENT TESTS When a test involves the use of test personnel or subjects in a hazardous environment, it is extremely important to fully instrument both the system and the people involved so that the test conductors can monitor in real time the exact status of the hardware and the individuals involved, such as an astronaut aboard a space vehicle or a pilot performing unusually hazardous maneuvers in an aircraft. Fail-safe communication links must be provided so that information links are not broken as a result of physical conditions or environmental anomalies.

Wherever practicable, rescue personnel must be made available for quick emergency response. In addition, system experts should be at hand to advise the test personnel aboard the system how to correct problems, how to eject or abandon the system, etc. All possible emergency events that can be anticipated should be analyzed, and procedures should be practiced before the test is begun. Above all, provide the test personnel as much onboard capability as possible for taking care of their own emergency; they may not have enough time to communicate with, and receive instructions from, the base test conductor.

Human Performance Measurement during Design Testing and Evaluation

As a general rule, human performance is measured in terms of how many errors a subject makes and/or how long it takes the subject to perform specific tasks. However, human subjects can also help judge the adequacy of a design either by rating the design against some set of standard criteria or by rating several designs against one another. The table below provides a brief description of several techniques for performance and/or design acceptance adequacy.

Method	Technique	Statistic	Application
Adjustment (average error)	The subject adjusts a stimulus until it is subjectively equal to or is in some relationship to a criterion.	Average of settings (average error of settings measures precision).	Absolute threshold; equality; equal intervals; equal ratios.
Minimal change (limits)	The experimenter varies the stimulus upward and/or downward. The subject signals its apparent relation to a criterion.	Average value of stimulus at transition point of subject's judgment.	All thresholds; equality.
Pair comparison	Stimuli are presented in pairs. Each stimulus is paired with each other stimulus. The subject indicates which of each pair is greater with respect to a given attribute.	Proportion of judgment calling one stimulus greater than another. Proportions sometimes are translated into scale values via assumption of normal distribution of judgments.	Order; equal intervals (under distribution assumption).
Constant stimuli	Several comparison stimuli are paired at random with a fixed standard. The subject indicates whether each comparison is greater or less than the standard. (This is a special case of the pair-comparison technique.)	Size of difference threshold equals stimulus distance between 50 and 75 percentage points on psychometric function.	All thresholds; equality; equal intervals; equal ratios.
Quantal	Various fixed increments are added to a standard, with no time interval between. Each increment is added several times in succession. The subject indicates the apparent presence or absence of the increment.	Size of sensory quantum equals distance between intercepts of rectilinear psychometric function.	Differential thresholds.
Order of merit	A group of stimuli, presented simultaneously, are set up in apparent rank order by subject.	Average or median rank assigned by subjects.	Order.
Rating scale	Each of a set of stimuli is given an absolute rating in terms of some attribute. The rating may be numerical or descriptive.	Average or median rating assigned by subjects.	Order; equal intervals; stimulus rating.

Organizing and Documenting Field Test Activities

The importance of documenting test results of hardware is well known, and such tests are recognized as absolutely necessary. Documenting how well people perform is equally important. Although this might appear to be an unnecessary admonition, too often it is believed that performance problems should merely be noted in a general way and that the actual performance of the people operating the equipment does not need to be measured. However, if a thorough program of human engineering has been pursued to this point, certain human performance standards will have been established. Therefore, it makes common sense to test the equipment user as well as the equipment.

As noted earlier, a final human-machine system test should examine not only the effectiveness of the hardware design but also the procedures, the supporting documentation (e.g., technical manuals), and the training.

Special test observation and recording forms should be devised to prompt field test observers and to provide them a place to record their observations. The accompanying sample format was created for the field testing of an aeronautical system. This test was oriented primarily toward field service and maintenance operations. Although only two of the test sheets are shown, they give an idea of what the field test observers were told to look for. Each type of test should have its own forms, tailored to fit the particular application and the features that will probably be observed.

HUMAN ENGINEERING FIELD EVALUATION FORM

TASK TIME RECORD:　　　　TASK: _____　　EVALUATOR_____

Preparation _____　　　_____　　DATE_____
Performance _____　　　_____
Clean Up _____　　　_____
Total _____　　　_____

TASK DIFFICULTY		APPARENT PROBLEM
1. Didn't know how to start.	☐	
2. Didn't know what to do next.	☐	
3. Missed a procedural step.	☐	
4. Got steps out of order.	☐	
5. Repeated step unnecessarily.	☐	
6. Initiated action too late; too soon.	☐	
7. Difficulty coordinating activity of two or more persons.	☐	

RECOMMENDED ACTIONS:

_____

PROCEDURAL PROBLEMS

HUMAN ENGINEERING FIELD EVALUATION FORM

TASK TIME RECORD:　　　　TASK:_____　　EVALUATOR_____

Preparation _____　　　_____　　DATE_____
Performance _____　　　_____
Clean Up _____　　　_____
Total _____　　　_____

TASK DIFFICULTY		APPARENT PROBLEM
1. Difficulty locating proper display or control.	☐	
2. Confused meaning of label.	☐	
3. Couldn't read display detail.	☐	
4. Mis-read display.	☐	
5. Missed warning light.	☐	
6. Difficulty positioning control.	☐	
7. Set control incorrectly.	☐	
8. Disturbed adjacent control.	☐	
9. Didn't recognize color code.	☐	

RECOMMENDED ACTION:

_____

DISPLAY-CONTROL PROBLEMS

Suggested Reading List

CAKIR, A., D. J. HART, and R. F. M. STEWART: *The VDT Manual,* Inca-Fieg Research Association, Washingtonplatz 1, D-6100 Darmstadt, Federal Republic of Germany, 1979.

Human Engineering Design Criteria for Military Systems, Equipment and Facilities, MIL-STD-1472 (latest revision).

LUXENBERG, H. R., AND RUDOLPH L. KUEHN (EDS.): *Display Systems Engineering,* McGraw-Hill Book Co., New York, N.Y., 1968.

Military Standardization Handbook: Human Factors Engineering Design for Army Material, MIL-HDBK-759.

Potential Health Hazards of Video Display Terminals; June 1981, U.S. Dept. of Health and Human Services, Nat'l institute for Occupational Safety and Health, Cincinnati, Ohio.

RUPP, BRUCE: *Human Factors of Workstations with Visual Displays,* International Business Machines Corp., San Jose, Calif., 1984.

VAN COTT, H. P., AND R. G. KINKADE (EDS.): *Human Engineering Guide to Equipment Design,* McGraw-Hill Book Co., New York, N.Y., 1972.

WOODSON, WESLEY E.: *Human Factors Design Handbook,* McGraw-Hill Book Co., New York, N.Y., 1981.

Index

About the Author

Wesley E. Woodson is President, Man Factors, Inc., a human factors research and consulting firm in San Diego, California. Experienced in both military and domestic system and product design, he has been an active designer of man-machine system concepts covering aerospace, man-in-space, architectural, highway and transportation, surface and subsurface, interiors, electro-mechanical, electronic, and business machines, and computer systems. He is the author of *Human Engineering Guide for Equipment Designers* in addition to over 200 technical reports and articles dealing with a wide variety of human factors research and application. In 1974 Mr. Woodson was the recipient of the Jack Kraft Award for significant efforts to extend application of human factors principles and methods to new areas of endeavor.